THE SEASONS OF

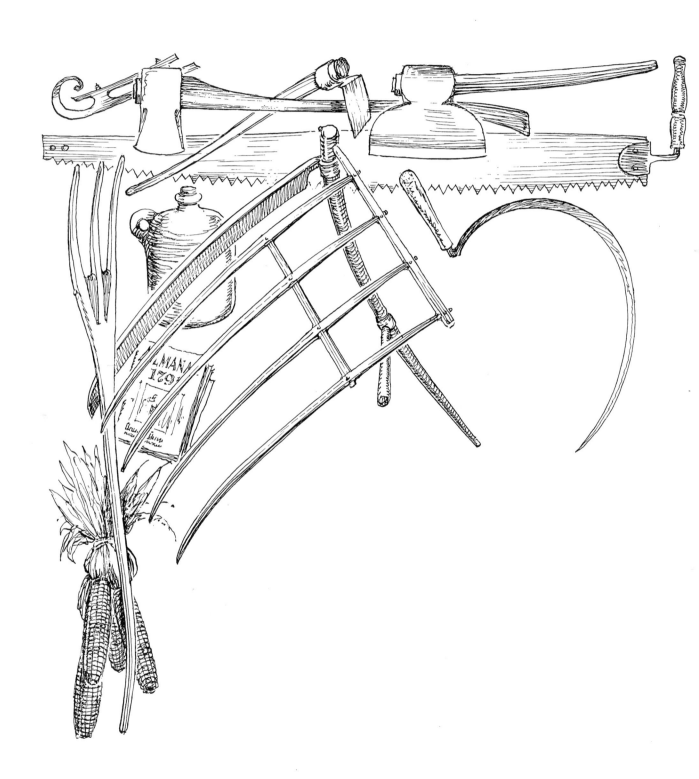

AMERICA PAST

by Eric Sloane

DOVER PUBLICATIONS, INC.
Mineola, New York

Bibliographical Note

This Dover edition, first published in 2005, is an unabridged republication of the work originally published in 1958 by Funk & Wagnalls, Inc., New York. The only significant alterations are as follows: The four full-color seasonal portraits are reproduced in color on the covers ("Autumn" on the front, "Spring" on the inside front, "Winter" on the inside back, and "Summer" on the back), and in black and white in their original positions. This necessitated moving the two original endpapers to their appropriate season inside the book: "Average dates of First Killing Frost in the Spring" now appears facing page 33; and "Average dates of First Killing Frost in the Fall" now appears facing page 83.

Library of Congress Cataloging-in-Publication Data

Sloane, Eric.
   The seasons of America past / by Eric Sloane.
      p. cm.
   Originally published: New York : W. Funk, [1958]
   ISBN 0-486-44220-9 (pbk.)
    1. Farm life—United States. 2. Agricultural implements—United States. 3. Seasons—United States. I. Title.

S521.S6249 2005
630—dc22

                        2004063467

Manufactured in the United States of America
Dover Publications, Inc., 31 East 2nd Street, Mineola, N.Y. 11501

Contents

Author's Note

Nostalgia, it has been said, is a great American disease. Yet an appraisal of the past need not be nostalgic. True nostalgia is "homesickness," and even the most ardent antiquarian would not so yearn for the past as to want to return completely. In this speeding world, the faster we travel, the farther back we leave our past. We soon find ourselves using all our powers to "keep up with things," and looking backward at all has become a lost art. Even beholding and evaluating the present becomes difficult.

We have actually come to believe today that we must either progress or retrogress. Each season of existence should be an entirely new one, according to twentieth-century thinking, and there is no such thing as intelligently remaining stationary. Next year's things, we assume, must necessarily be improvement on this year's, and to want anything but the newest, brands us as quaint.

Contentment too is considered a bogey in this century. Eugene O'Neill voices this modern opinion, saying, "One should be either sad or joyful. Contentment is a warm sty for eaters and sleepers." How different was America two centuries ago when Benjamin Franklin declared that "Contentment is the philosopher's stone that turns all it touches into gold!"

We often observe that great-grandfather had a knack of enjoying himself that we seem to have lost. It might be that his "seasons for fun" were more independent from his "seasons for work" than ours are today. It might be, too, that he devoted himself more completely to the moment.

That great American privilege and aim, the "pursuit of happiness," originally involved a now almost obsolete use of the word "happiness." Then, it meant "blessedness," or "a state of satisfaction or contentment," but now it suggests *fun*. The "pursuit of happiness" which we accept as an American heritage is, it seems, too often mistaken for a pursuit of fun. I am alarmed as I agree with Carl Sandburg that "Never was a generation . . . told by a more elaborate system of the printed word, billboards, newspapers, magazines, radio, television—to eat more, play more, have more *fun*." This, we are led to believe, is an American way, and a recipe for contentment. Yet the time for fun and the time for contentment were two very different seasons in great-grandfather's mind; and he fared fabulously well with both.

I am indeed grateful for the good things of this age, yet I feel there were certain things of the past which were good and unimprovable, many of which have become lost. It is both my lot and pleasure to look backward, to search the yesterdays for such carelessly discarded wealth. I am forever thankful for living at a time when many of the marks of early America still exist, before that fast-approaching time when they will all have disappeared into a far different landscape.

America, the richest nation in the world, has managed to be the most wasteful. We will be the first to admit this, and there is even pride in our voice. We spend our way into prosperity and out of recessions so that thrift is regarded a way of the past. Across our nation at present is written a record of land wastefulness never equaled in the history of the world. Land is "improved" by destroying it and building over the waste. We always forgive ourselves with the ready excuse that we can afford wastefulness. But there is always a reckoning, and even now we begin to wonder. We might wonder what other wasteful ways of everyday life have also become Americanisms.

The lost seasons of early America may sound like vanished trifles, but in a confused age when the most patriotic American must sometimes grope for words to explain his heritage, or to define "Americana," any material which contributes to a better understanding of our past is invaluable, and it is often the apparently small detail which contributes most.

The American heritage, as I see it, is grounded in the freedom and expression of the individual, and individual freedom, I maintain, was a fresher spirit a century or two ago. Individual expression was likewise richer. I believe that freedom becomes stale and expression becomes poor without constant appraisal.

In this age of "arms races" and "space conquest," the simple, basic philosophy of our past is too often ignored; and when the study of the past is mistaken for nostalgia, beware!

Eric Sloane
Weather Hill
New Milford, Conn.

Speeding up the Seasons

Blessed is he who takes comfort in seed time and harvest, setting the warfare of life to the Hymn of the Seasons.

J. W. HOWE

An early dictionary quotes the Bible in defining the word "season" as ". . . a time to every purpose." The complete quotation was an old favorite from Ecclesiastes, memorized by pioneer farmers, and quoted on children's samplers. "To every thing there is a season, and a time to every purpose under the heaven: A time to be born, and a time to die; a time to plant, and a time to pluck up that which is planted; A time to kill, and a time to heal; a time to break down, and a time to build up; A time to weep, and a time to laugh; a time to mourn, and a time to dance; A time to cast away stones, and a time to gather stones together; a time to embrace, and a time to refrain from embracing; A time to get, and a time to lose; a time to keep, and a time to cast away; A time to rend, and a time to sew; a time to keep silence, and a time to speak; A time to love, and a time to hate; a time of war, and a time of peace."

After "Season: a *time* to every *purpose*," the same dictionary adds, almost like an afterthought, ". . . also one of the Quarters of the Year." Noah Webster's first dictionary completely left out the division of the seasonal year, simply saying that a season is "a fit time." Before the mid-1800's, the word "season" had become an Americanism denoting the intricate timings of nature, and people did not automatically associate it with spring, summer, fall, and winter. Seasons were timed to the vagaries of weather, the appearances of the moon, the peculiarities of growing things, or the rise of an occasion. They were the slow heartbeat of the American countryside. They were the countryman's calendar.

Rainfall proper for planting or transplanting was at one time called a "seasonal fall" of rain. Timber cut at the time of sap-run appropriate for the wood's adaptability to its destined use was called "season-cut," and those who believed that the moon's orbit influenced the rise and fall of sap, timed tree-felling not only to the proper "season of the moon" but even to the exact hour.

Illness suffered during the first few months of settling in a new country (usually caused by a change in drinking water) was called "seasoning fever." Roof-shingles were put in place, floors were laid, seeds were planted and grain reaped all in their own special seasons. Early American life was one continuous parade of seasons.

Twentieth-century progress has eliminated the old ways, and the seasons of America past have vanished, but accounts in ancient farm ledgers and the methods of the old-timers still echo them loudly. Without realizing a complete lack of logic, we have long been explaining old-time thoroughness as the work of people "who had the time." Yet it is *we* who really have the most time: the old-timer's chores before breakfast alone often equaled a whole day's worth of work today. The craftsman's workday was once ended automatically by the setting of the sun; how foolish of us, with our time-savers, to credit *him* with "more time!"

Time, it seems, need not be "saved" at all, unless there is something to do with it. Yet time-saving has become such an American obsession that many of us now spend much of our life collecting and paying for our time-savers. How strange, too, that we who strive to live a long life also choose to live it quickly. Life, after all, cannot be speeded up any more than music can without becoming strained and finally grotesque. In all the books about farming before the mid-1800's, almost no words may be found about speeding-up methods. Volumes were written, however, about the value of "taking proper time." People had a righteous contempt for anything that was rushed.

The old-time farmer said, "The fast-grown punkin usually turns out to be the 'pore' one," and the saying still goes. Despite all the progress of science, the average man of today most often ends up physically poorer than the same man of a century or two ago. Life-expectancy in youth and middle age, for example, has naturally risen because of medical progress; but because he lacks proper exercise, nutrients, and mental relaxation, the man over *sixty* today is actually a weaker man than his ancestor was at the same age. *The elderly man of today has less chance of living than did his counterpart of the past!* In 1832 a special census was taken of all people in the United States over one hundred years old. The possibility of inaccuracies was taken into consideration and hearsay, such as reports about Negro slaves, was ruled out. It was found that at least one person in every forty-five hundred Americans was a hundred years old or more. Today the figure is only one in thirty-four thousand.

Few of us can comprehend it, but two things have vitally changed the character of food and mankind during the past century. One is that much of present-day food is grown from a more depleted soil and it is therefore proportionately deficient. Another is that nutritional losses from *refining* and *processing* further lessen the value of food. In some cases the loss may be as much as fifty percent.

Very much like today's food, which is grown larger and prettier but with less nutrient, the average man of today is larger, but lacking in stamina. A striking thing about ancient armor and early American clothing is its smallness. Statistics show that man eats more now, but gets less food value than he did in the past. He becomes softer and flabbier as he increases in size.

Spring, summer, fall, and winter seem almost to vanish today; we ice-skate on artificial ice, eat strawberries or watermelon all year round, and go about our work regardless of the weather or the time of year. Timber is cut during all seasons now, and lumber is dried in an instant. Fruit is forced and tinted to appear tree-ripened; even hens are made to lay faster by illuminating chicken barns throughout the night. Almost nothing need be waited for any more. Certainly there must be a loss involved with such speed—a loss of appreciation and enthusiasm for things which can be had too easily.

Chemists in 1957 gave honors to a fourteen-year-old girl for speeding the three-minute egg. By adding glycerine to water, she found, a three-minute egg can be boiled in one-and-a-half minutes. It was significant that no honors were given to the scientist who the same year discovered that slowly boiled eggs are more digestible than those rapidly cooked. Any results of speeding up nature, it seems, are hailed as twentieth-century progress. The eventual and total results might be like those described in a 1778 almanac. "Overplanted fields," it says, "make a rich father but a poor son. Every bit taken from the earth must be returned, or it suffers."

A century later another almanac ironically reminded us that "the United States which founded its civilization on ten inches of top-soil has now lost *one-third* of it! Rushing the seasons and forcing the land is surely aging the soil toward a premature death."

The vanishing art of waiting is constantly discouraged nowadays. Thoreau, who insisted that "all good abides with him who waiteth wisely," seems to be speaking in a faraway voice from a distant and plodding world. The smartest man, we are too often led to believe, is the fastest man or the one who can make a "snap decision." The wise man, however, is he who takes time appropriate to the decision. "If it's a quick decision you want," the old-timer always insisted, "the answer is no!"

Rushing the seasons of life can sometimes be dangerous, even fatal. Synthetic sex hormones have for the last few years been given regularly to farm animals to accelerate their growth, although the cancer-producing tendencies of estrogen on animals is well established. A cancer-inciting drug is used at present to cause sped-up growth in fowl. Marketable chickens have contained many times the amount of this drug which sufficed as a daily dose to induce cancer in mice. Over thirty million chickens are inoculated yearly with such chemicals to increase their weight, and at least half the nation's beef receives the same treatment. Stilbestrol, the chemical usually involved, is regarded by scientists as biological dynamite that is not destroyed by cooking.

When a super-market operator was asked why he sells oranges colored with harmful "ripe orange color" dye, he replied that: "People just haven't time for ripening green oranges; dye makes oranges look better, too. The average American buyer buys by appearance rather than by analysis, anyway."

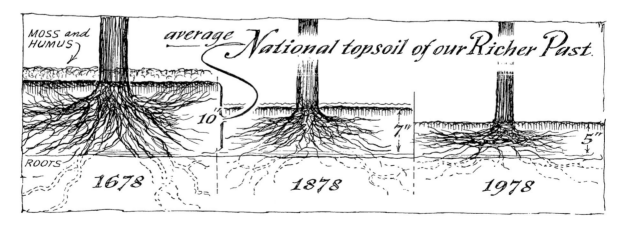

MOSS and HUMUS

average National topsoil of our Richer Past.

10"

7"

5"

ROOTS

1678   1878   1978

Twentieth-century progress is sometimes more involved with speed and quantity than with quality. For the sake of speed and reduced costs, modern manufacturers have become artists at imitating nature. Some ice cream manufacturers use piperonal (excellent for killing lice) as a substitute vanilla flavoring. Diethyl glucol (paint-remover) and anti-freeze material is used as an egg flavor. Ethyl acetate once was used for cleaning leather and textiles, but it makes a good artificial pineapple flavor now. Oil paint solvent, amyl acetate, fakes banana flavoring; benzyl acetate imitates strawberry; butyraldehyde fakes nut flavor. Some food manufacturers guarantee "artificial flavors absolutely pure."

Another sad thing about unnaturally rapid progress is that it so often results in actual *regression*. When any group progresses as a whole, there becomes less *apparent* need for individual intelligence. By having others think for us and design our work and pleasure, we now live comfortably without certain knowledges that only a century ago were essential. In losing our need to know so many things, our list of general knowledge has at last become exceedingly small; it equips us well for business, less well for the sciences, and very poorly for living the full life.

Being streamlined to make the journey of life with "just enough" baggage might help us "arrive" quicker, but we risk finding ourselves unequipped and empty when we get there. This might explain why the intelligent person of two centuries ago was so often more accomplished than the educated man of today. It also serves to explain the fuller enthusiasm for life and reverence for the cycles of nature that was so much a part of the past. In the days of almanacs, when man lived close to the earth and the sky dictated every move, there was a beautiful and profound understanding of time and life and the spiritual connection between the two.

The speeding-up of life and the eventual "squeezing out" of nature's seasons means much more than the mere elimination of minor pleasures like the strawberry season: it eventually affects civilization. The orderly maturing of man, for example, is changing; there is less age demarcation now than there was only a half-century ago. Television entertainment, like most other modern Americana, is designed more and more for appeal to both children and grown-ups; the divisions between childhood and adult life are already doubtful in many fields. Grandmothers in shorts are no longer grotesque; silver hair is no longer a badge of dignity to be worn proudly. The wonderful childhood days of mumbly-peg and swimming holes or just plain day-dreaming are going and proper playtime is finally something planned and supervised by "child-experts."

Growing from childhood to adulthood was once an unhastened process, either by chores on the farm or by apprenticeship in the city. It was a properly slow progression. Now the passing from school days to the adult business world is a quick and often unprepared step.

The modern child's seasons of growing up are crowded in such a manner that they often disappear completely. In 1957, the president of one of New York's top advertising agencies surprised his own industry by saying in a television talk, "Children a generation ago were lucky to see one comic cartoon a week . . . at the Saturday afternoon movie. Today's youngsters choose from a dozen . . . morning, noon, and night. Television today is telescoping into the space of a few years the entertainment interests that once extended over a lifetime. Are we using up our interest so fast that boredom sets in with abnormal and destructive swiftness?"

A typical 1958 Sunday television-broadcast day featured stories with over two dozen separate murders; some of the murders were of simple cowboy species; others were involved with more bizarre crime settings. How strange that the modern world seldom notices the grotesqueness of such timing. If a sermonette on crime should come on to interrupt our Sunday murder series every little while, we would object violently and brand the idea as "corny" or out of place; yet crime entertainment on the Sabbath doesn't seem out of place to most of us. The old art of adhering to a time for every purpose lent a rare dignity to everyday life.

The circle of nature's seasons and the discipline of seasonal routine were once signals for individual activity, but the seasons of today involve group activity. With the addition of competition and spectator sports, we now specialize in group sitting and group watching. The benefits of watching a three-hour baseball game on television or at the ball park, can never equal a few minutes of play at the game yourself. Theodore Roosevelt, who was one of the last of America's great individualists, said, "It is far more important that a man should play something himself, even if he plays it badly, than that he should go to see someone else play it well."

Buying maple syrup at the super-market during National Maple Syrup Week will never bring as much joy of self-expression as tapping a maple tree and boiling the sap yourself. Of course we haven't the time nowa-

Forgotten Toys
from the Seasons of
Childhood Games .. do you remember ..

Hoops

Stilts

One·a·Cat

1. HIT PUCK HERE,
2. PUCK RISES,
3. ,,BAT IT AS YOU WOULD A BASEBALL.

PIECE OF WOOD.

APPLE OR POTATO

ROD

STRING

STICKS

DEVIL

The Diávalo
PLAYED SINGLY
OR AS A DOUBLE "CATCH-GAME".

Apple **Ball**
OBJECT IS ⸻ DISTANCE.
THE LONGER and SPRINGIER THE ROD,
THE GREATER THE POWER and SPEED.

The Whip·top
WAS STARTED BY THE
HANDS AND WHIPPED
INTO SPEED BY A CORD,
LEATHER THONG OR AN
EEL-SKIN !

(REMEMBER
TOP-SEASON ?)

Snow·Scooter
A REAL CHALLENGE

Grass sled

barrel Stave

hog's head staves.

17

days to grow our own food or to make our own clothing and household furniture, but we must expect some starvation of self-expression. There is a moral emptiness that comes when we are deprived of the satisfaction of doing things for ourselves.

Of course, those four seasons of the year will always be with us, and we are still properly conscious of them. Yet we are already far less aware of spring, summer, fall, and winter than we used to be. Every schoolboy once knew the arc of the winter and summer sun, the meaning of equinox, and all the signs of nature that revolve in harmony with each season. Many people today believe that the sun sets in the true west all year round, instead of from the southwest to the northwest according to the season. All early farmhouses were planned and built in accordance with the seasonal arc of the sun and the seasonal course of winds, but houses are now built to conform with street plans. Only recently, architects "discovered" the advantage of building according to the seasonal sun, and they call it a modern idea.

With no intention of insulting the reader's intelligence, the drawings on the following page explain our four annual seasons. The drawings show the great variation of daylight and sun position during spring, summer, fall, and winter. You will find most early American farmhouses planned around this knowledge, taking advantage of whatever the seasons of the year have to offer.

In this strange world of forgotten knowledge and progress which so often defeats itself, even the slightest look into the past is revealing. It might be inspiring to consider the obsolete ways of men who geared their living to time instead of trying to beat it, and linked their days with nature without always trying to conquer it. As we watch the old-timers enter their various seasons of work we might realize that competition is not the only path to progress. We might learn about the mystic ingredient which so often gave results of craftsmanship and integrity beyond present-day achievement. We might even revive the fast-disappearing custom of allotting "a time to every purpose" and "to everything a season."

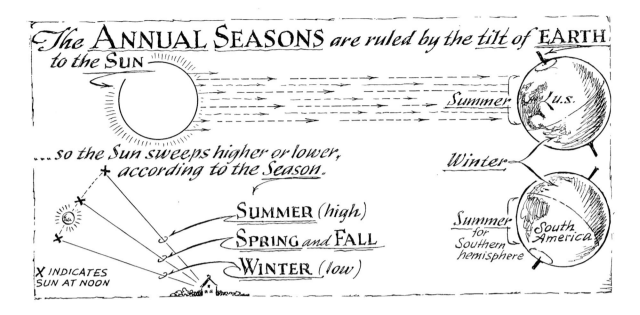

The ANNUAL SEASONS are ruled by the tilt of EARTH to the Sun

Summer in U.S.

Winter

Summer for Southern hemisphere — South America

... so the Sun sweeps higher or lower, according to the Season.

SUMMER (high)

SPRING and FALL

WINTER (low)

X INDICATES SUN AT NOON

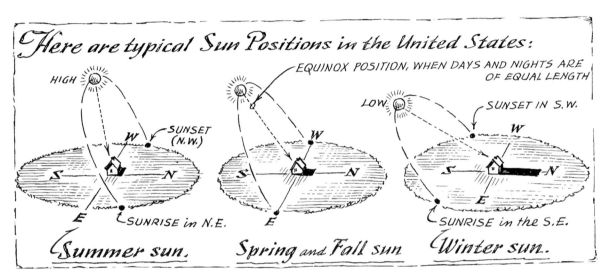

Here are typical Sun Positions in the United States:

HIGH

EQUINOX POSITION, WHEN DAYS AND NIGHTS ARE OF EQUAL LENGTH

LOW

SUNSET IN S.W.

SUNSET (N.W.)

SUNRISE in N.E.

SUNRISE in the S.E.

Summer sun.   Spring and Fall sun   Winter sun.

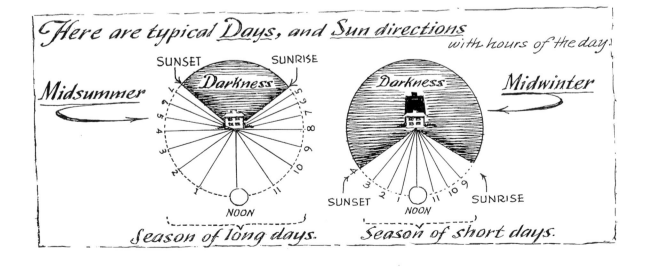

Here are typical Days, and Sun directions

with hours of the day

SUNSET   SUNRISE

Darkness

Midsummer

Darkness

Midwinter

NOON

SUNSET   NOON   SUNRISE

Season of long days.   Season of short days.

Agrarian Kindergarten

Colonization — Farming — Industry

1650 1700 1750 1800 1850 1900 1950

The first farmer was the first man;
all historic nobility rests on the
possession and use of land.

RALPH WALDO EMERSON

It would have been impossible for the American colonists to have landed elsewhere on this side of the Atlantic and find so little in nature unfamiliar to them. From Massachusetts to Virginia, nearly all plants and animals were closely related to those overseas. The greatest difference in the New World was its seasons. Not only was the weather lustier, it was entirely unpredictable. The season for planting one year was far from being the proper season the next year. Summers were hotter and winters were colder than those in the same latitude of Europe. If any new set of rules for farming and husbandry were to be devised, it was fitting that it be designed around the seasons of America.

It seems strange that the American colonists should have needed schooling in agriculture, yet for the first century their extent of farming knowledge came largely from the Indians. The first settlers came in search of furs or naval stores or lumber, and for almost every reason but to become farmers. The fact that they needed above all else to sustain themselves by products of agriculture was realized slowly.

Many think of the early Plymouth colony as being a miniature farming settlement but there were at first not even farm animals. The first few cows brought over died from being fed poor grass. In 1650 there were fewer than two hundred plows in all of North America and most of them were in Virginia. The ancient wooden plow was useless in most rocky New England soil and for many years there were only thirty-seven plows in the whole colony of Massachusetts Bay. Most farmers were

20

compelled to break up bushes with their bare hands and prepare the ground with mattocks as crude as those used by the Indians. With a wealth of providence in the ground and within the forest, a whole century slipped by before American farming really began. Even at the end of that century, the average American farmer could carry his complete stock of farm implements (except his cart and harrow) upon his own back. For many years the American colonist had depended upon England for grain and hay to keep himself and his cattle from starving.

In the mid-1700's the American farmer, possibly the true American, was born. Learning agriculture from scratch, the new-world farmer went about it in a grand and becoming manner, thoroughly and with great dignity. The rules and philosophy that evolved during this graduation made farming a creed and a way of life that was as American as the Declaration of Independence.

The first American *Farmer's Manual* began with the words, "Venerate

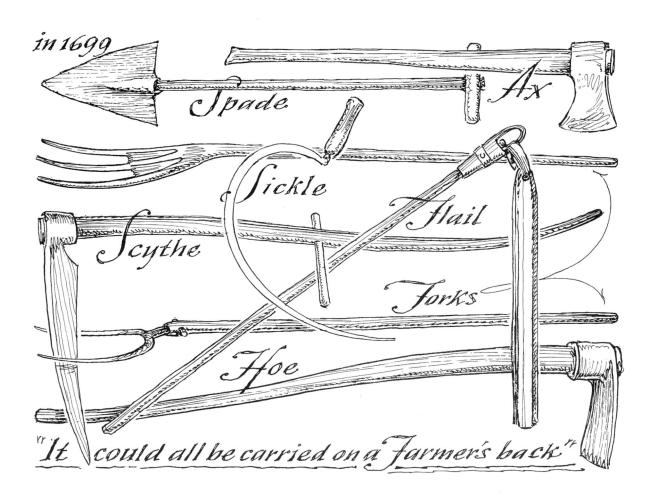

in 1699

Spade · Ax · Sickle · Flail · Scythe · Forks · Hoe

"It could all be carried on a Farmer's back"

the plow! Husbandry was the first employment of man, therefore the most ancient, the most honorable, and, above all, of divine appointment." Farming books treated farming as a philosophy or religion, not an industry, as we do now. The farmer enjoyed economic independence, and agriculture was recognized to be the fundamental employment of man. Simply being a good farmer in those days automatically made you a good American.

In 1958, President Dwight D. Eisenhower summed up American progress by saying, "We have progressed from an isolated farm economy to a world industrial economy." It sounds sophomoric in this advanced world to quote a tribute to early American agrarianism, but these words of our first President are worth pondering: "Let the merchants boast of their nice calculations, their stocks in business well laid in," said Washington, "and contemplate their profits, amounting to vast stores of wealth, in expectancy; the success of all their schemes and even their own personal support, depends upon the farmer." This is as economically true today as it was then.

The following bit is from one of the first handbooks on American farming. Its moral philosophy, which permeates even the technical instructions of the book, explains the original "American Way" better than any study of the mechanics of modern competitive industry could now explain it. Spiritual references in a technical handbook seem awkward today, but we must remember that the world of two centuries ago was not a world of business; it was a world of farming and religion, the companion pursuits of everyday life.

> Farmers, I have before remarked, ye are the lords of this lower creation, and I have shown you this by clear demonstration; reverence yourselves, by your industry, economy, temperance, sobriety and punctuality; with all the Christian virtues; and you will compel the world to reverence you. Should any one order, rank, station, or individual in society, withhold from you the tribute of respect, justly due to your rank, and worth in society, let him alone; reflection will correct his error. Let no advantages of improving your knowledge in the science of husbandry, escape your attention; apply this improvement in knowledge, to the improvement of your farms, by little and little, as circumstances, and your means may afford you opportunity; a well directed industry, with the blessing of God, will enable you to

surmount all difficulties, and will make you both rich and independent, and your families after you. . . . Under such husbandry, the merchant will flourish; the artist, and the labourer will flourish; the agriculture and commerce of our country (those handmaids of nations) will flourish; our country will become the garden of the world, and America the store-house of the world.

The early American farmer has been accused of ignorance and superstition, yet when we truly study him we find an extraordinarily learned man. The mathematical and scientific aspects of early farm textbooks that were so much a part of the primitive American picture were often above the cultural intelligence of today's average farmer.

When farming became a business instead of a religion of living, a revolutionary change occurred in American thinking. The farmer who farmed just for himself was looked upon as a strange and obsolete American, and his methods became rapidly outmoded.

One ridiculed practice of early farming is the use of "moon seasons." With our modern lighting equipment, we could not be expected to understand why a farmer would plan to sled his produce to market "during ye Gibbous Moon of January." Yet anyone of a century or two ago knew that the best sledding weather, and nearly two weeks of moonlight, were most possible at that particular time of January. Many of the old "moon dates" and "moon seasons" were no more than those periods when nature worked with you rather than against you. Drovers, for instance, made good use of moonlight; farmers who wanted their cattle driven to market often planned ahead to a full-moon date for the trip. In the old days, eyes were more generally accustomed to the dark, and many farm chores were done entirely at night. "Moon time" was no more mysterious than clock time, and moon dates were used as commonly as our calendar dates are now. The calendar itself, after all, is based upon the moon's phases, although we seldom stop to realize it.

It is true that folklore surrounds many of the old "moon seasons," and although many modern scientists ridicule moon effects, there are also reputable scientists who accept them. "The moon has such potential influence in the various parts of her orbits," says the *New England Farmer*, "that by cutting one tree three hours before the new moon and

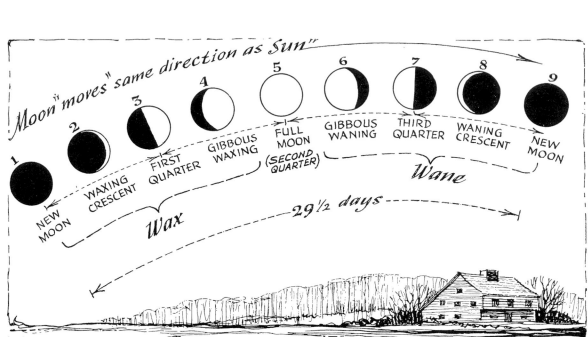

These are the PHASES of the MOON and their names.

another of the same kind of tree six hours afterwards, a difference in the soundness of the wood is noticeable."

The season for cutting wood depended upon what the wood was to be used for, the kind of tree, and mainly upon what the sap was doing within the tree at the time. It was also the opinion of the old-timer that as the moon affects the mechanics of all liquids on earth just as it does the tides of the oceans, so does the moon affect the flow of sap within living plants. The drawings of the moon's phases for one complete cycle (close to a month) show the sequence of lunar appearances and the proper names given to them. Strangely enough, even in this day of approaching moon exploration, the average person does not even know the moon phases. We often call the slender wisp of an early moon a "new moon," not knowing that a true "new moon" is actually without light at all. We call the first and third quarters of the moon "half moon," often without even knowing whether it is waning or waxing. Few modern school children know what a gibbous moon is, and most of us are even surprised to learn that the moon follows about the same sky route as the sun.

The moon was not only a timepiece and calendar, but a guide for those who traveled by night. Everyone knew that a line drawn through

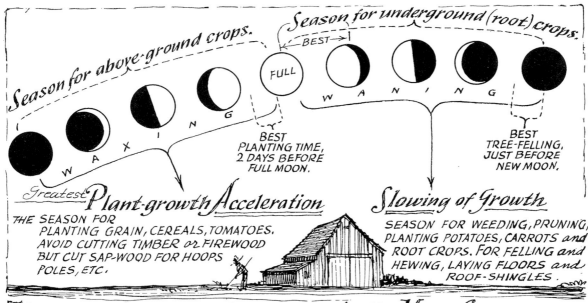

Season for above-ground crops.

Season for underground (root) crops.

BEST →

FULL

W A X I N G

W A N I N G

BEST PLANTING TIME, 2 DAYS BEFORE FULL MOON.

BEST TREE-FELLING, JUST BEFORE NEW MOON.

Greatest Plant-growth Acceleration

THE SEASON FOR PLANTING GRAIN, CEREALS, TOMATOES. AVOID CUTTING TIMBER or FIREWOOD BUT CUT SAP-WOOD FOR HOOPS POLES, ETC.

Slowing of Growth

SEASON FOR WEEDING, PRUNING, PLANTING POTATOES, CARROTS and ROOT CROPS. FOR FELLING and HEWING, LAYING FLOORS and ROOF-SHINGLES.

These are things that were done in the Moon Season era.

the horns of the crescent moon would generally indicate north and south; that a waxing crescent followed closely behind the sun setting in the west, while the waning crescent rose in an easterly sky, just ahead of the sun. The full moon is always directly opposite the sun, so again you can orient yourself. Looking at the early almanacs, you might be overwhelmed by the abundance of astronomical data. None of this was beyond the comprehension of the simplest farmer, who was, first of all, sky-minded.

The right-hand section of the drawing shows things that the early farmer did during both the waning and the waxing of the moon. This book does not argue that the chart is entirely correct, but it does accept that the results were strangely beyond present-day accomplishments. Year-round cut wood, for example, warps or rots quickly, and there are many other similar examples which indicate that moon lore had a scientific basis.

The early English countryman's bible was Thomas Tusser's almanac-like list of things to do and when to do them, titled "Five Hundred Points of Good Husbandry." A collection of hints for farm success, this book was a gem of do-it-yourself wisdom. Many of its rhymed pages had a whimsical humor not unlike Yankee drollery, while others contained gems of the Ogden Nash-type of simplification, such as:

Keep oxen from cowe
For reasons yknow.

Tusser instructed the countryman in the proper manner of doing everything from building a barn to determining the best time to retire or rise. Such lines as the following made his book a favorite among the American students of farming:

In winter at nine, and in summer at ten,
To bed after supper, both maidens and men.
In winter at five the servants arise,
In summer at four it is very good guise.

and

A doore without locke is bait for a knave,
A locke without key is a fool that will have.
One key to two lockes if it breake is a griefe,
Two keys to a locke in the end is a thiefe.

In many early American houses, the Bible and Tusser's manual comprised the complete library. The title page and a sample text page reproduced here were from a copy found boxed with a Bible in a cupboard in an early Connecticut farmhouse. It is odd that Thomas Tusser's name should be forgotten today; perhaps it is because his rules were made for the old-world seasons. Perhaps the American almanac filled the new need better. Although Thomas Tusser's philosophy was the basis for much of the almanac's traditional format, his name was forgotten immediately with the publication of the first American almanacs.

One of the many myths surrounding Benjamin Franklin is that he was the founder of the American almanac. The earliest American almanac was the first book to come off the College Press at Harvard, the *Almanack for 1639*. Five years before the birth of *Poor Richard's Almanack*, James Franklin (Benjamin's brother) was printing the *Rhode Island Almanack*. And three years earlier than that Nathaniel Ames in Boston was publishing an almanac as witty and wise as Poor Richard's

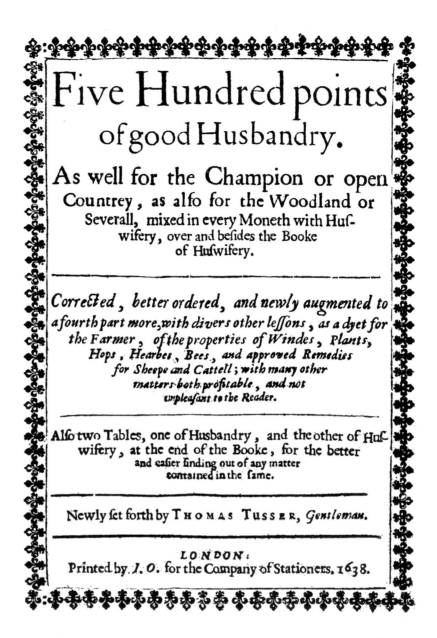

Five Hundred points
of good Husbandry.

As well for the Champion or open Countrey, as alfo for the Woodland or Severall, mixed in every Moneth with Hufwifery, over and befides the Booke of Hufwifery.

Corrected, better ordered, and newly augmented to a fourth part more, with divers other leffons, as a dyet for the Farmer, of the properties of Windes, Plants, Hops, Hearbes, Bees, and approved Remedies for Sheepe and Cattell; with many other matters both profitable, and not unpleafant to the Reader.

Alfo two Tables, one of Husbandry, and the other of Hufwifery, at the end of the Booke, for the better and eafier finding out of any matter contained in the fame.

Newly fet forth by THOMAS TUSSER, *Gentleman.*

LONDON:
Printed by *I. O.* for the Company of Stationers. 1638.

was to be—a book that is regarded by most experts as being "better than Franklin's," and that reached an annual circulation of sixty thousand copies.

People seldom realize the need of the almanac in the old farmstead. Almanacs were the timetables to the general seasons of early American life; they were as valuable as an accurate timepiece. Let us assume that farmer Brown of 1750 finds that his clock has wound out and stopped. How can he reset it? He cannot carry the stopped clock to his nearest neighbor, who is twenty miles away. With his almanac as a guide, how-

such slivers doe well, for to lie by the wall.

2 Git grindstone and whetstone, for tooles that is dull
   or often be letted, and fretted belly full :
A wheelebarrow also, be ready to have,
   at hand of thy servant, thy compasse to save.

Grinding stone, and whetstone.

3 Give cattle their fodder, in plot dry and warme,
   and count them for myring, and other like harme :
Yong colts with thy vennels, together goe serve,
   lest lurched by others, they happen to sterve.

4 The racke is commended for saving of doong,
   so set as the old cannot mischiefe the yong :
In tempest (the wind being northly, or east)
   warme barth under hedge is a succour to beast.

5 The housing of cattell, while winter doth hold,
   is good for all such as are feeble and old :
It saveth much compasse, and many a sleepe,
   and spareth the pasture, for walke of thy sheepe.

Housing of cattell.

6 For charges so little, much quiet is woone,
   if strongly and handsomely all things be done :
But use to untackle them, once in a day,
   to rub and to lick them, to drinke and to play.

Get trusty to tend them, not lubberly squire,
   that all the day long, hath his nose at the fire :
Nor trust unto children, poore cattell to feed,
   but such as be able to helpe at a need.

8 Serve Rie-straw out first, then wheat straw & pease
   than Oat-straw and barly, then hay if ye please :
But serve them with hay, while the straw stover last,
   then love they no straw, they had rather to fast.

9 Pokes, forkes, and such other, let Barly spie out,
   and gather the same, as he walketh about :
And after at leasure, let this be his hire,
   to bath them and trim, at home by the fire.

Forkes and yokes.

10 Is well at the full of the moone, as the change,

           D   Sea

ever, the *sun* and *moon* and *tide* served as constant and accurate time-pieces. He therefore set his clock by the pages of his almanac. It was never a vague book of superstition; rather was it a scientifically accurate and useful book. As a *long range* weather forecaster, it had no less value than any device we have today, according to box-score comparisons.

The difference between old-world and new-world seasons largely involved the fact that major temperate zone air-masses flow toward the east. Air that blows across England and coastal Europe has first been tempered by the Atlantic; it has also been influenced by the Gulf Stream, which ends its curved ocean voyage directly off the British Isles. It is milder, less turbulent air after a whole ocean has cooled it during summer and warmed it during winter. Air over the Atlantic coast of America, however, is mostly dry, cold, continental air, fresh from and close to its source in the cold dry northwest. The quality of such a climate is one of abrupt seasonal changes and sudden temperature drops. New England, for instance, being in the direct path of all polar air-masses, is a prime theater for rapid weather changes. An average of two cold air-masses move across the New England countryside each week, causing almost bi-weekly showers. The roller-coaster temperature variations of New England are anything but ideal for large scale farming, and the almost unpredictable frost seasons will forever harass gardeners. Yet if a New Englander were asked what territory has a climate

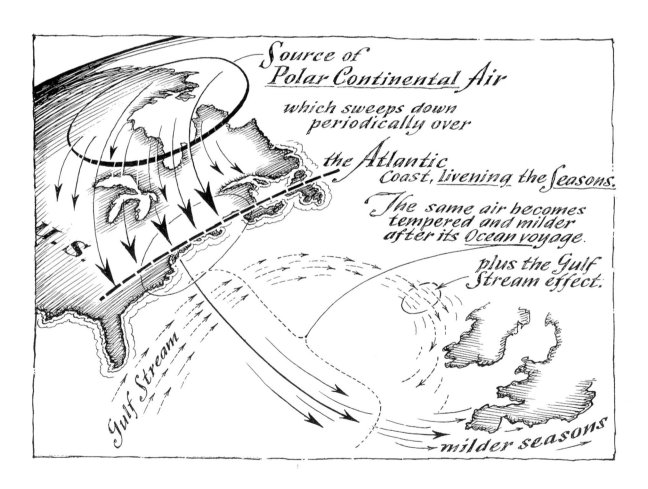

Source of *Polar Continental Air* which sweeps down periodically over the *Atlantic* Coast, livening the *seasons*. *The same air becomes* tempered and milder after its *Ocean voyage*. plus the *Gulf Stream effect*.

Gulf Stream

milder seasons

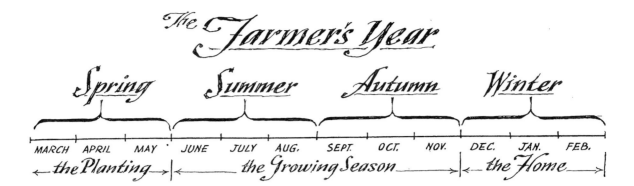

The Farmer's Year

Spring   Summer   Autumn   Winter

MARCH   APRIL   MAY   JUNE   JULY   AUG.   SEPT.   OCT.   NOV.   DEC.   JAN.   FEB.

←—the Planting—→|←———the Growing Season———→|←—the Home—→|

ranging from a hundred and ten degrees above to fifty degrees below zero, he would very likely not recognize it as his own.

Early American weather extremes are usually regarded as folklore exaggerations, but there is much evidence that such extremes did exist. One such case concerns the "year without a summer season" in 1816, when many places in New England and the middle states saw snow during every month of the whole year. There is a good first-hand account of this cold summer of 1816 in the "Autobiography of Chauncey Jerome" of Bristol, Connecticut, pioneer American clock maker. He writes:

> The next Summer was the cold one of 1816, which none of the old people will ever forget, and which many of the young have heard a great deal about. There was ice and snow every month of the year. I well remember on the 7th of June, while on my way to work (dressed throughout with thick woolen clothes and an overcoat), my hands got so cold I was obliged to lay down my tools and put on a pair of mittens which I had in my pocket. It snowed about an hour that day. On the 10th of June my wife brought in some clothes that had been spread on the ground the night before which had been frozen stiff as in Winter. On the Fourth of July, I saw several men pitching quoits in the middle of the day with thick overcoats on, and the sun shining bright at the same time. A body could not feel very patriotic in such weather. I often saw men, when hoeing corn, stop at the end of a row and get in the sun by a fence to warm themselves. Only half enough corn ripened that year to furnish seed for the next.

The New World was well called the "Place of Many Seasons." Spring, summer, fall, and winter lessen in contrast as you go south-

30

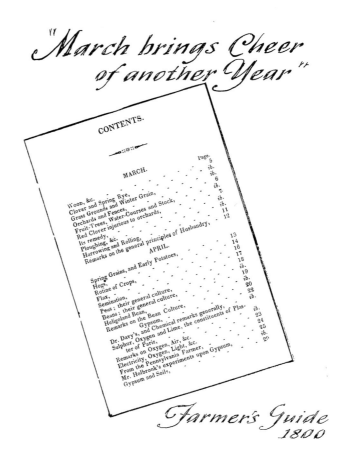

"March brings Cheer of another Year"

Farmer's Guide
1800

ward, and their differences disappear completely off the southern tip of Florida, but the center of the world's most pronounced seasonal effects is on a line drawn horizontally through the middle of the northern states.

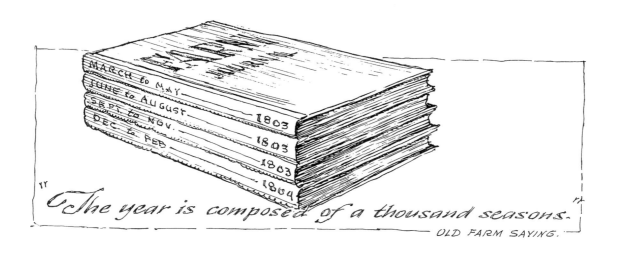

"The year is composed of a thousand seasons."
— OLD FARM SAYING.

In the city, man starts the new year on the first of January, but nature's year begins more appropriately in the spring. At one time, all farm calendars and diaries, almanacs and agricultural manuals began with March, for the old-time year began on March 25. In 1752, about a century and a half after Pope Gregory XIII had corrected the early calendar, his corrections were finally adopted by England. Eleven days were dropped from the calendar at that time, and New Year's Day was changed from March 25 to January first. For a while the American colonists celebrated both old-style and new-style New Year's Day, but then grudgingly they gave up the old calendar. But personal farm records remained "old style," with the annual routine of farm seasons beginning where it should, in the spring. "The new year is at our door," reads a typical rural diary, "spring is with us in March when we are yet sitting by the fireside . . . now in the sky, now in the waters, with returning birds, upon some single tree, in a solitary plant, and each milder touch gives pleasure to those who are content to await the natural order of things."

So the chronology of this book, like that of old diaries and farm ledgers, begins with March.

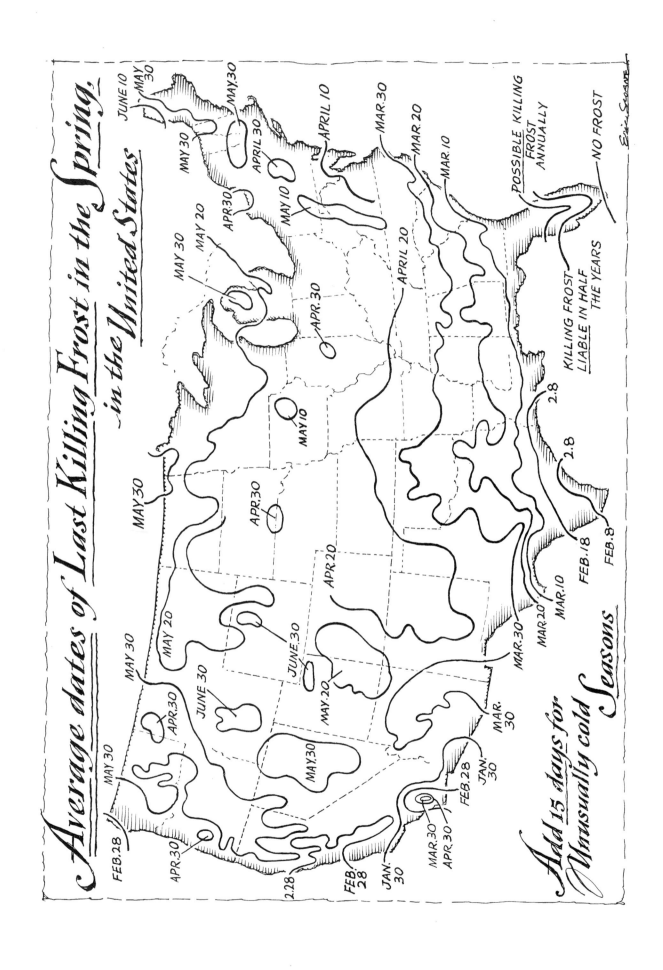

Average dates of Last Killing Frost in the Spring, in the United States

Add 15 days for Unusually cold Seasons

POSSIBLE KILLING FROST ANNUALLY

KILLING FROST LIABLE IN HALF THE YEARS

NO FROST

March

Now Sol o'ercomes old Winter's frowns,
And soon unlocks the frozen grounds;
So drink March out with Season'd mead,
To flavor every New Year deed.

OLD ALMANAC

The New Year drink is nothing new, yet its evolution is complicated. Any season for toasting the future calls for strong spirits, but the old-style New Year's Day was in March, and the old-style March drink for the occasion was *mead*. At the back of this book you will find recipes for making mead, along with various other Americana information, all listed alphabetically. Mead is a drink made from honey; what with the thousand and one liqueurs that adorn the backbars of our modern drinking places, it is strange that the oldest ceremonial drink of all (or at least some form of it) is entirely absent.

We might wonder where the word "honeymoon" came from; it originally meant "honey-month" or "mead-month." For one full moon season (from moon-phase to the next like moon-phase, or about a month), it was once the custom of newly married people to celebrate with mead. This season was called the "honey-moon season" and it still is, mead or not.

The early American farmer, who drank cider daily at his table instead of water or milk, was nevertheless a sober man. But mead and "hardened cider brandy" were always in order, no matter what the after-effects, during the March preparations for the coming seasons of labor. Not very different, it seems, from our present New Year's Day custom.

The typically American doughnut, once a pre-Lenten cake on the farm, became strangely mixed up with the drinking of cider on New Year's Day. Shaped like a quoit, the doughnut was designed to be tossed into the air for children to catch on that gay day before Lent known as Fat Tuesday, Doughnut Tuesday, Shrove Tuesday, Carnival Day, or Mardi Gras.

When we fry doughnuts in deep fat we seldom think of the operation as being symbolic of Fat Tuesday (a literal translation of the French *Mardi Gras*); nor do we think of the word "carnival" as meaning the end of meat-eating (from *caro*, flesh, and *levare*, to take away).

In appraising the future of a farm, fences were reckoned a prime necessity. Almanac after almanac starts the month of March with, "Look to your fences." Fences are now of minor importance, but a century or two ago, when split-rail fencing around a farm was often worth more than the land itself with its farmhouse and barns, things were quite different. Such advertisements as, "Farm for sale, completely fenced (includes building and barns)" were not unusual. Even in 1850, the fencing for a three-hundred-acre farm cost about five thousand dollars at the current price level. At the beginning of the 1800's, when much of America's virgin woodland was stripped of trees, even firewood became a valuable part of real estate, and it was a common thing for a farmer to will his property or sell his farm "along with a good supply of firewood."

March in mountainous areas was often mid-winterish in spirit, delaying the start of the actual farming seasons; blustery March, then, was the ideal season for storing up firewood and splitting fence-rails or, as the farmer put it, "puttering with wood." March winds dry out the winter-cut logs in the woods, making them easier to haul in and "The difference in saving between green and dry wood," says the 1821 *Farmer's Almanac*, "will pay the expense of sledding, besides the extra trouble of kindling fires. Both are worthy of attention."

Although March was the month for hauling in and cutting up wood, the actual felling of trees for fence material was often done during the second running of sap, in August. "Even the twigs then," says the *Pennsylvania Farm Journal* for 1850, "will remain sound for years." In the 1700's it was the practice to cut fencing during winter, but with the 1800's came the knowledge that virgin oak, hickory, and chestnut split into rails in winter would last only twenty years, while the same trees felled in August lasted fifty. (For today's inferior second-or-third-growth timber you may divide these figures by three; there is actually that much difference.)

Many famous pictures have been painted of Abraham Lincoln splitting rails with an ax. Only a few were historically accurate, showing him with a gigantic wooden mallet or "beetle" and wooden wedges or "gluts." Rails were always split by hammering on them with wedges, never by striking them with an ax. Mauls and wedges once so constituted the woodchopper's main tool supply that it became an American colloquialism to refer to a man's complete possessions as his "maul and wedges," a phrase that one never hears now. Gluts were made of hard wood made even harder by exposure to a slow, even heat. The old-

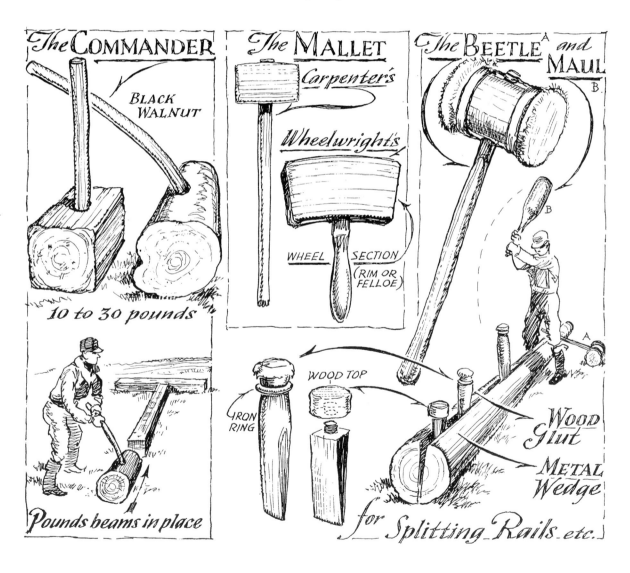

time fireplace was seldom without a number of gluts being "slow-seasoned" near the fire.

The use of wooden hammers is now almost a lost art, but the workshop of a century ago had a great variety of them. Antique dealers sometimes exhibit gigantic wooden hammers, giving people the idea that men of the past were some sort of giant. Some of these hammers, known as "commanders," are nearly too heavy to lift, the head being a two-foot length of black walnut log. But these were swung below the hips, their efficiency being in their weight rather than velocity of impact. It was the comparatively lighter "beetles" that were swung overhead like an ax.

Timber cut at the proper season, or dried in the proper season and split at the proper season, is so easily cleaved with a wooden hammer and wedge that the work offers profound satisfaction and is peculiarly fascinating. Abraham Lincoln knew this relaxing pleasure, saying that some of his "best thinking was done when working hardest at splitting rails." One blow, applied with the precision of diamond-cutting, will split a rail just to your liking if you are expert enough.

No early American season was more definite than sugaring time. The right time is usually between mid-March and mid-April, when the sap is flowing properly. Then the nights are cold enough to freeze sharply and the days warm enough to thaw freely. A cold northerly or westerly wind assures this phenomenon, while a warm easterly or southerly wind precludes it. The thermometer must not rise above forty degrees by day, nor sink below twenty-four degrees at night. It is this magic see-sawing between winter and spring that decides the sugaring season. The sun seems almost to draw the sap up and the frost to draw it down, for an excess of either stops the flow, as if by magic.

Let John Burroughs describe the art of maple sugaring:

> Maple sugar in its perfection is rarely seen, perhaps never seen, in the market. When made in large quantities and indifferently, it is dark and coarse; but when made in small quantities—that is, quickly from the first run of sap and properly treated—it has a wild delicacy of flavor that no other sweet can match. What you smell in freshly cut maple wood, or taste in the blossom of the tree, is in it. It is then, indeed, the distilled essence of the tree. Made into syrup, it is white and clear as clover honey; and

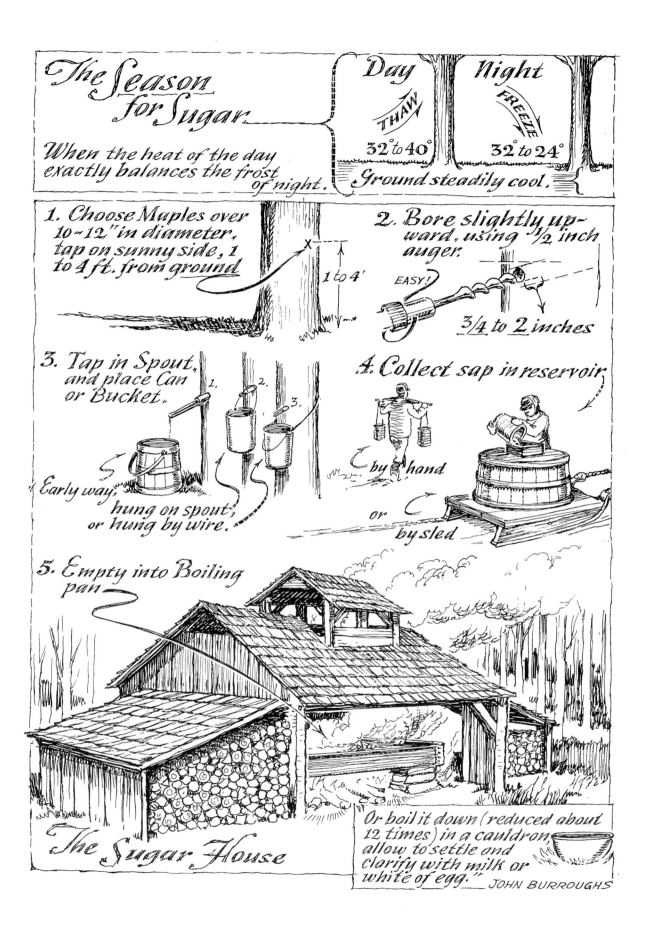

The Season for Sugar

When the heat of the day exactly balances the frost of night.

Day	Night
THAW	FREEZE
32° to 40°	32° to 24°

Ground steadily cool.

1. Choose Maples over 10~12" in diameter, tap on sunny side, 1 to 4 ft. from ground

1 to 4'

2. Bore slightly upward, using 1/2 inch auger.

EASY!

3/4 to 2 inches

3. Tap in Spout, and place Can or Bucket.

1. 2. 3.

Early way, hung on spout, or hung by wire.

4. Collect sap in reservoir

by hand

or

by sled

5. Empty into Boiling pan

The Sugar House

Or boil it down (reduced about 12 times) in a cauldron, allow to settle and clarify with milk or white of egg." — JOHN BURROUGHS

crystallized into sugar, it is pure as the wax. The way to attain this result is to evaporate the sap under cover in an enameled kettle; when reduced about twelve times, allow it to settle half a day or more; then clarify with milk or the white of an egg. The product is virgin syrup, or sugar worthy the table of the gods.

The art was the discovery of the American Indians, but it didn't take the white American long to acquire a taste for sugar. At present, through refining and overconsumption, doctors agree that there is nothing more detrimental to the national diet than sugar. Digested starch is the major source of sugar for animal metabolism, and ample starch for the purpose is provided by grains and vegetables in the diet. But the typical American fare of refined bread, cereals, macaroni, cake, and pie deluges the system with sugar. And to top it all, we pour refined sugar on top of everything that will take it. The days when all sweetening was done with blackstrap molasses, honey or maple sugar were richer than are these days of abundant sugar with its nutrient values refined away. "For myself," says the *Farmer's Almanac* of 1818 for March, "I have done using sugar, and feel better for it. But those who will still use this luxury, as I shall call it, had better be attending to their maple trees."

Possibly, breeding could have done for the maple tree what has been done for the sugar beet and sugar cane, to make its yield greater. But the slow-growing maple is not adapted to the speed of modern farming; it takes forty to fifty years before a tree is in full production. The experimentation, selection, breeding, and grafting done with any common fruit tree has never been afforded the American maple. Its sugar has been taken for two centuries without benefit of any research into cultivation, yet with no loss to the trees. Despite our tremendous increase in population, two hundred years ago there was four times the amount of maple sugar and syrup produced each year as there is now. Sugaring was hard work, but the American farmer made such a cheerful season of it that the whole family looked forward to sugaring, making it more play than work.

The trick of successful sugaring is to have everything ready for the very short season. A great deal of the enjoyment comes from the preparation. On winter nights before sugar time, the men would whittle sumac spiles at the fireplace; metal rods were heated in the coals to be used for burning through the pithy centers to complete the spout. Preparing

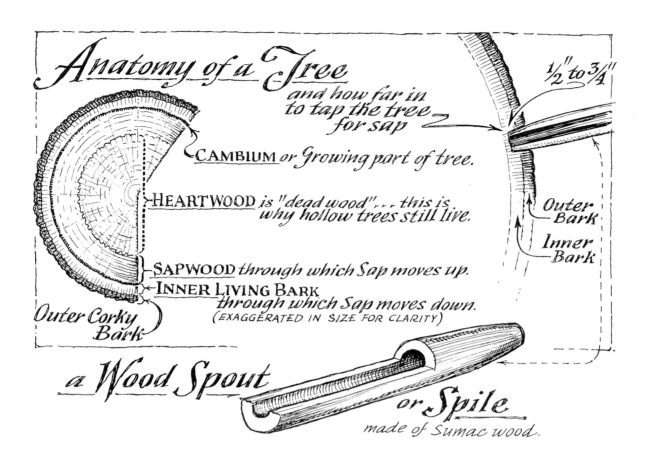

Anatomy of a Tree
and how far in to tap the tree for sap

½" to ¾"

CAMBIUM or Growing part of tree.

HEARTWOOD is "dead wood"... this is why hollow trees still live.

Outer Bark

Inner Bark

SAPWOOD through which Sap moves up.
INNER LIVING BARK through which Sap moves down.
(EXAGGERATED IN SIZE FOR CLARITY)

Outer Corky Bark

a Wood Spout or Spile
made of Sumac wood.

four or five hundred spiles was ample work for any junior craftsman. The making of maple sugar buckets was not usually a fireside industry, but was left for the cooper to do. Once made, they lasted a lifetime. Notice in the drawing how the hoop (put on wet) lapped over itself and dried hard and tight. Not one nail was used in these old buckets, which· could outlast our modern metal pails.

A covered sugar bucket was also made for storing loose sugar which, like modern sugar, often hardened in the containers. The *sugar-devil*, a tool that confounds many an antiquarian, was a device used for loosening hard sugar. Small sugar-devils were often made on the farm, but larger ones were later manufactured for the cane sugar barrels of the old country store.

Maple sugaring is a short-seasoned business, and it doesn't usually involve the cheap labor that southern cane sugar has available, so it is not favored by the interest of big industry. As a diminishing luxury business, it is getting smaller on the national scale each year. At present the actual

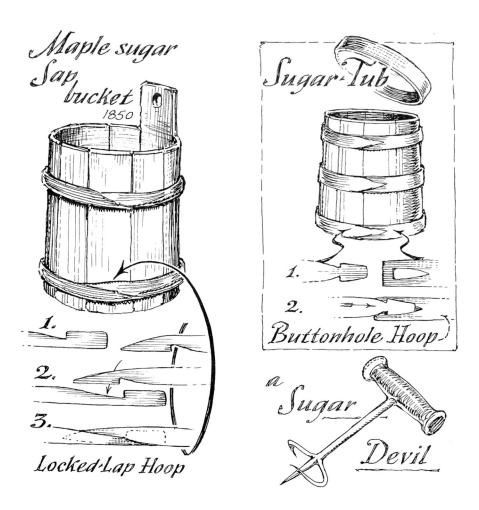

Maple sugar
Sap
bucket
1850

1.
2.
3.
Locked·Lap Hoop

Sugar·Tub

1.
2.
Buttonhole Hoop

a Sugar
Devil

labor cost alone of one gallon of maple syrup is over a dollar and a half; the full cost is about two and a half dollars. A third of all the maple sugar we now produce is from New York State, and only thirty-seven percent comes from New England. In 1860 maple syrup production ran to more than six and a half million gallons; at present we produce probably half that, or less. Although it is one of our greatest natural heritages, the business of making maple sugar is already considered a quaint pursuit, another example of modern speed and quantity winning over yesterday's quality.

The March chore of laying up new fuel wood heralded the end of winter, the season of the hearth. Hearth lore is persistent, and the most nostalgic of all echoes of the past. It is interesting how the sight and smell of a wood fire gives one a feeling of firmness and security. Man builds electrical and air-conditioned homes, yet includes in them a wood-burning fireplace as a symbol of spiritual comfort. If he can't do that, he often builds a dummy one as a symbol of graciousness and repose in a home that otherwise lacks it.

Winter months spent around the early farmhouse's only source of heat, the open fireplace, were crowded ones. The fireplace's enormous size, it seemed, was never quite adequate, because everyone in the family had some peculiar claim to the heat and light closest to the hearth. Besides heating and cooking equipment, there were always a few pieces of wood present, being seasoned by the winter fire. Special wood for ax handles and other farm tools was laboriously dried at the fireplace, and even lightly charred for strength. Special pieces were often left near the fireplace for as long as a year, to render them properly seasoned.

Strange recipes for wood seasoning were handed down from generation to generation without any other reason than that they always seemed to work. Burying wood in wheat, salt, dry earth, or hot ashes were among the favorite methods. One, that of soaking the tool-handle in stable manure for a month before smoke-drying it in the fireplace, was possibly least popular with the cook.

Seasoning an ax handle on the spit-hooks

March, not long ago, was the season for colds, and starting with the popular old sulphur-and-molasses precaution, the parade of spring remedies was endless. We shake our heads in shame at yesterday's patent medicines, but our expenditures for "tired blood" and headache cures now would make great-grandfather's patent medicine bill look puny. The common cold was not, it seems, as prevalent in the farming days of the 1700's as it became during the 1800's and 1900's, although the earlier days were much wetter and more rigorous ones. The first cold remedies were simple enough, and are still worth trying, inasmuch as we have nothing better. George Washington's cold remedy was to eat a toasted onion before going to bed. (Others may be found in the back of this book under *cold remedies*.) Only fifty years ago the mustard-growing business thrived on that first treatment for colds, the hot mustard foot-bath.

There are few people who want to remember this good old treatment because the procedure is a lengthy one. It takes time, and it involves going directly to bed with a hot lemonade drink. Pills now eliminate all this fuss; they might one day eliminate the common cold, but to date are no improvement on the old remedies.

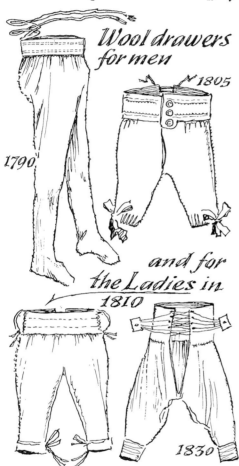

Wool drawers for men 1790 1805

and for the Ladies in 1810 1830

One of the causes of colds in the 1800's was the change from wintertime's heavy underwear. Long drawers were adopted in about 1790 and were at that time designed to completely envelop the feet. They were first called "draw-ons" and they sported a long drawstring which was wrapped twice around the waist. Women's long underdrawers (no relation to pantaloons) came in about 1810 and reached only to the calf. Once having put "longies" on, no sane person dared change to light-weight undergarments before the first of June. The flannels of a century ago were as heavy as a modern topcoat; they were full of all the tickling and scratching impurities of homespun wool. "If you fell out of a tree and caught on your old fashioned long underwear," it was said, "you were saved!"

A tombstone in Jonesport, Maine, reads: "Here lies the body of Ephraim Daniels, Who chose the dangerous month of March to change his winter flannels."

April

When April blows his horn*
It's good for hay and corn.

* Thunder. An old saying is that an early thunderstorm heralds a good farm season.

Kite season and marble season started simultaneously, yet neither had an exact date. As soon as the weather was right, kites and marbles appeared magically as if from nowhere and became the occupations of all small boys. Early American marbles were made of baked clay, and no small challenge of the game was the erratic route that the not-too-round spheres took over the ground. There were three ways of shooting a marble: *trolling,* or throwing the marble along the ground underhand, like a softball; *hoisting,* in which the player knelt and flipped a marble from knee level; *knuckling,* in which the player shot with the middle knuckle of his forefinger touching the ground.

Kite-flying was an art a century ago, and kite fairs were popular meets, usually held in April, where one could win prizes for flying height, lifting power, or workmanship. Benjamin Franklin's most famous kite was not his lightning-experiment job at all, but the one he used to pull him while swimming. The idea took hold to such an extent that swimmers had "kite races," floating on their backs and being pulled over a water course by their kites. Kites were also used in cable or chain bridge-building to carry over the first string, which, of course, was reenforced by heavier material until it became rope or cable. "When the bridge at Niagara was built," reads an 1849 account, "a single wire was carried over by a kite and on that wire was taken over a cable." As used to be the custom, a prize of

43

five dollars was offered to the first boy who could send his kite to the opposite bank, and Homan Walsh, who won the Niagara Bridge award, enjoyed the fame till the end of his days.

April is the month of spring noises. After the winter's quiet, the first hum of a spring night sounds thunderous. The first peepers of the northern states promise one freeze after their debut, but by the end of April their chorus is joined by tree toads and insects of every description. A century ago, when the American countryside was fifty percent wetter, it is estimated that the insect and frog population was more than twice that of today. The April chorus must have sung an impressive lullaby to great-great-grandfather after his long day's work.

"April," says the 1850 almanac, "is here. Your wood is now housed and piled, and your sleds are stored safe for next winter. This most important month for the farmer is the first season for *manuring*."

Manure has become a word strangely changed from its original use. Never a choice subject in polite conversation, it now has reference more often to filth than anything else. Coming from the word root meaning "hand," however, manure was first a verb meaning "to cultivate." To "manure the land" meant to work it by hand.

In the 1700's the word manure was used to signify any material at all that was added by hand to the soil. The *Christian Almanack* defined manure as "bones, woolen rags, fish, leather, soap suds, brine, lime, mud from the swamps and ponds, refuse hay, ferns, rotten wood, shells, ashes, flesh scraps, dung and urine of all kind. . . ." A richer mess could hardly be imagined, but the miracle of compost is the clean sweet soil, fragrant as a spring bouquet, that results. The nature of compost involves the most important chemistry of human life and, left alone to decompose naturally, the vital soil that results is one of the great riches of our earthly heritage.

Nowadays we often think we are being generous with our synthetic fertilizers, but nearly half of all farm work not long ago was the business of returning to the soil each year what had been taken from it the year before. "Manure," says the *Farmer's Almanac* of 1818, "manure is the thing. What: do you imagine you can eat your cake and have it, too? This cannot be. Therefore you must plaster your mowlands with manure." The farmer of the past, if he could return, would be more impressed with our extravagance in destroying garbage than he would with many of our farm improvements.

An early Massachusetts agricultural report says, "We use from four to ten cords of manure, and average eight cords per acre, on our fields." It seems odd that manure should be measured by the cord, as we now associate cord measurement only with firewood. The manure sled, which was four feet wide, four feet high, and eight feet long, was commonly known as a cord-sled (or, with wheels, a cord-wagon). This sled or wagon measure, we might presume, was an obscure beginning of our modern cord measurement for wood.

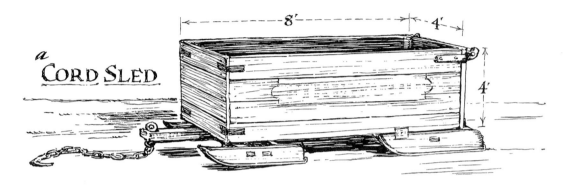

a CORD SLED

It was the early custom to put manure *beneath* root seeds (such as potatoes, onions, carrots, etc.) and *on top* of the above-ground crops. Potatoes planted in April (during the proper phase of the moon) were often planted upon a layer of plaster-of-Paris manure and topped, strangely, by an oval stone. The use of a stone as "manure" is often regarded as an Irish superstition, but stones do actually enrich the earth nearest them. One of the earliest writings on American farming suggested that scattered heaps of cleared stones be left in the field for several years before being piled into fences. Crops were often grown around such stone-heaps for as long as five years. Removed finally, "the lande where the stones were, will sprout three times the crop than all the land around about." Present-day landowners finding strange piles of stones throughout their woods, believe them to be Indian graves; but they are merely abandoned "manure stones" that never reached the stone wall stage. A Yankee farmer in Maine, advertising his land for sale, facetiously listed it as "guaranteed to raise a fine crop of stones." But the Irishman who bought it for a potato farm regarded the stony land as offering fine permanent manure for his potatoes.

First April... Second April... Third April.

"Clearing Bee... Stone manure piled. Fence building.

A group of Irish Presbyterians brought our first potatoes into Boston in 1718. They originated in South America, but became known as the poor man's food in Ireland, where they first won the title of "Irish potatoes." New Englanders would have little to do with potatoes at first because of the superstition that they shortened a man's life. Another rumor was that they were a powerful sex stimulant. Even after such superstitions wore off and potatoes were not only accepted but became the second crop of the average American farm, they were still used mostly for feeding cattle. Over nine thousand bushels of them were sent from Philadelphia in 1747 and sold as cattle feed, and as late as a century afterward the *Farmer's Manual* suggested that "potatoes be grown near the hog-pens as a convenience toward feeding the hogs."

Toward the end of April, and after the potatoes were in, came the proper season for clearing the back lands of stumps and stones. Winter frosts had heaved them to the surface, and the first spring thaws had loosened the stones' stubborn purchase in the ground. April was the time to get out the stoneboat and "twitch the big ones out" if they were to be moved at all. Some farmers dug holes alongside the bigger stones, rolled them into the holes, and covered them up; this explains many a big stone found only a foot or two beneath the surface on old farms.

Stoneboats are flat, runnerless sleds made to slide over the earth. In the days of poor roads they could not only be maneuvered around the countryside better than wagons, but they automatically repaired the roads (or made brand new ones) as they went along. Toll-roads actually wel-

46

Stone Boats were the pick-up trucks of yesterday.

...sometimes driven by reins like a wagon.

3 inch Oak planks

comed stoneboats, yet often either refused to accept very narrow-wheeled wagons or charged extra toll for them. The "boats" scraped and smoothed the road, whereas narrow wheels made the worst ruts. It was not uncommon for a farmer to drive to town and back again by stone-boat instead of by wagon. There are still remote places of unimproved roads where such vehicles outnumber wagons.

Only if one could realize the size of the trees that once covered the American countryside and the completeness of clearing done before 1875 could one appreciate the immensity of the vanished business of stump-pulling. In 1850, America was dotted with virgin pine stumps so durable that they might have lasted for several generations. "The American landscape is sprinkled with stumps," said an English writer, "like freckles on a farmer's face. Some dig them out and use them as ugly stump fences, but others leave them in place, where they will probably stay for the next century, and grow crops all around them."

There were about five hundred kinds of stump-pullers on the market in the 1800's, and stump-pulling became a common American profession. Two men with a yoke of oxen and a stump-puller could travel indefinitely across the country (while the ground was not frozen) and never run out of work. Stump-pulling was one of the few cash businesses, and at twenty-five cents a stump, the standard price in 1850, a man could pull from twenty to fifty stumps a day and make a most exceptional living for those days.

Stump-pullers and stoneboats are now outmoded by the bulldozer for

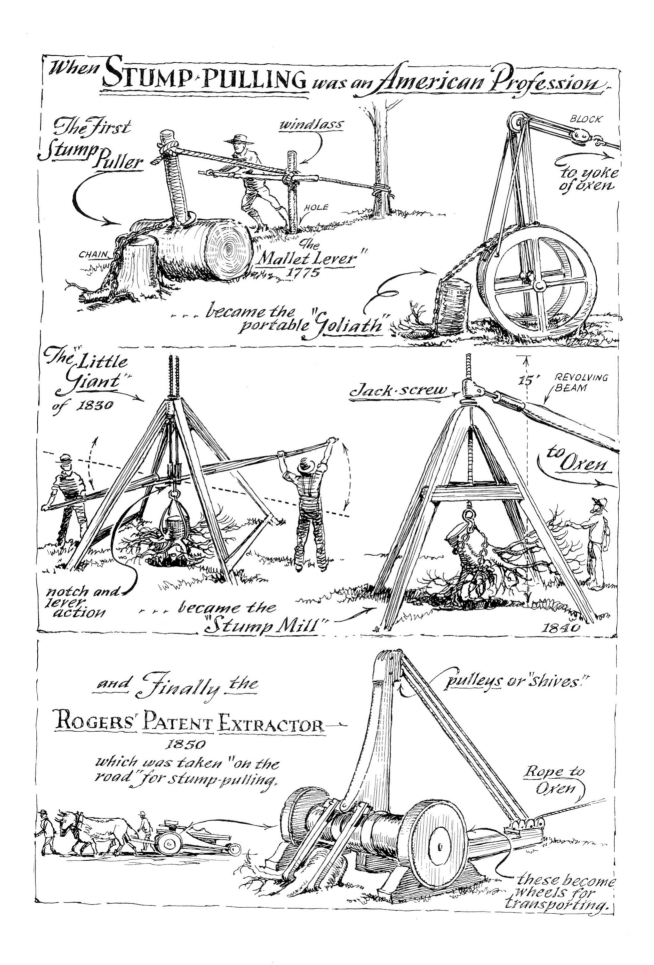

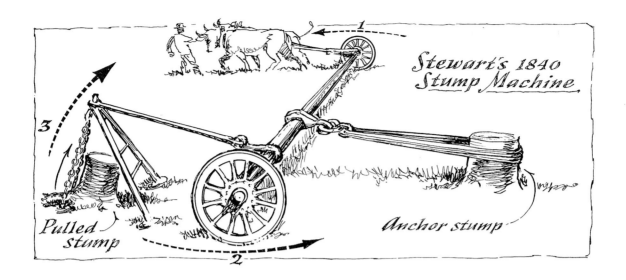

Stewart's 1840 Stump Machine

Pulled stump

Anchor stump

land-clearing, but to the real lover of the land bulldozing is an unpleasant sight. The undignified way in which the machine removes topsoil and changes the pleasing contours that nature took centuries to develop is a disheartening spectacle.

Stones, which were once regarded as symbols of permanence, were valuable for foundations and walls. Cement blocks do a quicker job to-day, so the ancient art of stone-masonry is a fast-vanishing occupation. And natural stone is forever pleasing to look upon. There seems to be a tumbling back of years and a reverence for the past in setting up a stone wall, that leaves man feeling better for having made it. The countryman accepted his crop of stone and developed for it a feeling of warmth and workman's pride that is hardly understandable today.

That stones should disappear from the landscape is incredible, yet nowadays landscapers go hundreds of miles to find suitable specimens for rock gardens. You may even buy "artificial stones for lining your driveway" from the mail order houses!

May

A Plow on a Field Arable Is the Most Honorable of Ancient Arms.

ABRAHAM COWLEY

In the country, May is a month of smells. To the city person, a "pleasant odor" is usually associated with flowers, but to the countryman it is more often the smell of the soil. After breathing the lived-in air of the farmhouse all winter, it is good to inhale the new smell of outdoors and the clean body of earth. This is one of the things that makes spring plowing, that hardest of farm chores, one of the most refreshing.

May is the season of the plow. It is the time for turning up wintered soil to meet the spring sky. "In May," say the old almanacs, "your Indian corn must be planted; this is the basic chore and first field work of the year." From the first turning of the soil to the Johnnycake and Indian pudding on the table, corn embodies the spirit of the early American farm more than anything else.

There will probably never be final proof that corn originated in America. The first colonists accepted corn as American because it was not only strange to them, but was never mentioned in the Bible.* It was, moreover, one of the few things the Greeks had no word for. Actually the word *corn* describes only the kernel, and there has never been a definite name for the plant that we now call corn. Derived from Old English, the word *corn* meant the grain of wheat, oats, or any other kernel-bearing grass. The first colonists named the strange new grains "Indian corns." Although what we usually call Indian corn today is the

* Corn is mentioned in some English translations of the Bible, but only as a synonym for grain.

5 0

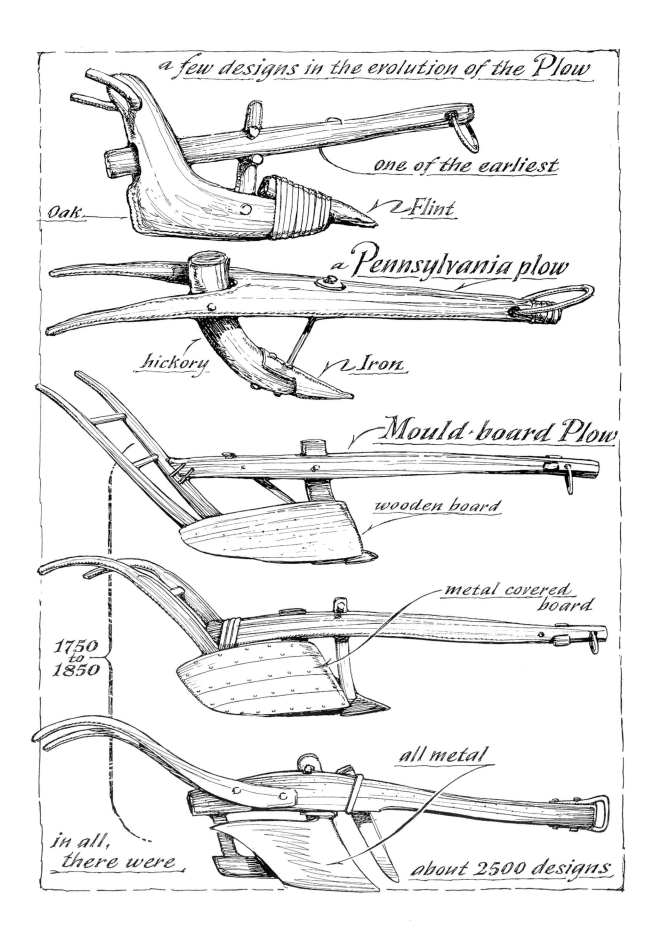

a *few* designs in the evolution of the *Plow*

one of the earliest

Oak

Flint

a *Pennsylvania plow*

hickory

Iron

Mould·board Plow

wooden board

metal covered board

1750 to 1850

all metal

in all, there were

about 2500 designs

decorative multicolored ear, all cob-type corn is Indian corn and was called that until a few years ago. Few recognize corn as a well-developed seed of grass. *Corn* in England refers to wheat; in Scotland and Ireland the word refers to oats; in the United States and Australia to the cob-seed or Indian corn.

Corn exists only with man, for without him to plant it, it would disappear! Its seeds are of such quantity and quality that when a whole ear falls to the ground, none of the seeds will reach the reproductive stage; most will die quickly from the too strenuous competition for nutrients and moisture. Only by man's proper planting of a *few* seeds in a proper hill will corn survive.

In 1492, during Columbus' explorations in Cuba, natives were found growing a "sort of grain they call maiz which was well tasted, bak'd, dry'd and made into flour." Early Indian corn and maize being the same thing, this was probably the first mention of corn in New World history. Where the American Indians had learned the secret of raising and cross-breeding corn is a mystery, but then the beginning of corn itself is a mystery. Like the mule, which is a cross-breed of ass and horse, hybrid corn is stronger and larger because of that characteristic of nature which brings forth greater strength and size by cross-breeding than by inbreeding. So the early name for American hybrid corn was "mule-corn."

Corn whiskey, that American institution of the liquor world, was being made in Kentucky as early as 1776, and within two decades a corn whiskey mellowed by charred oak was perfected in Bourbon County, Kentucky. Legend has it that the process was discovered accidentally when lightning struck the barn of a Bourbon County farmer who had a white-oak barrel of whiskey stored beneath the flooring. One of the results of this misfortune was a phenomenon—the barrel was charred, but its contents were improved. So "bourbon whiskey" was born!

May was once the season for sending "May baskets," now a forgotten custom. The first spring flowers were gathered by young girls and left in baskets on the doorstep for their parents, especially on the first of May. Perhaps some enterprising florist will revive that custom with a "National May Basket Week," which would certainly add more charm to the modern scene than does nonsense like "national hot-dog week" and such.

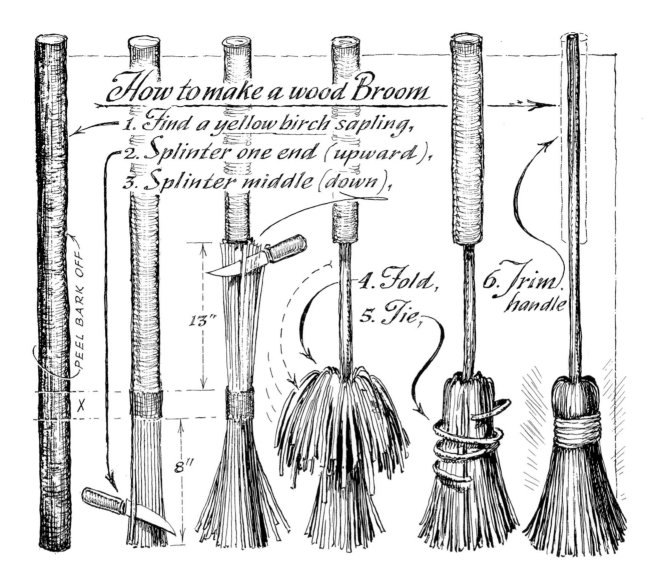

How to make a wood Broom
1. *Find a yellow birch sapling,*
2. *Splinter one end (upward),*
3. *Splinter middle (down),*

PEEL BARK OFF

X

13"

8"

4. *Fold,*
5. *Tie,*

6. *Trim handle*

Maypoles are relics of the English past, but the American farmer knew another "maypole." May was the season when he went into the woodlot to collect "pole wood." Although there is little use for them nowadays, poles were once an important part of farm equipment. Used as slides and rollers for moving heavy loads, as movable hay-floor beams and hay stack supports, or for hanging tobacco, they were used wherever wood was expected to have unusual give-and-take.

Poles are most efficient if they are cut in May; then ash and hickory contain their highest percentage of oil. During this time the growing bark is spreading, making it easier to remove; in another month it will have tightened and hardened. Such woodlore was not only the knowl-

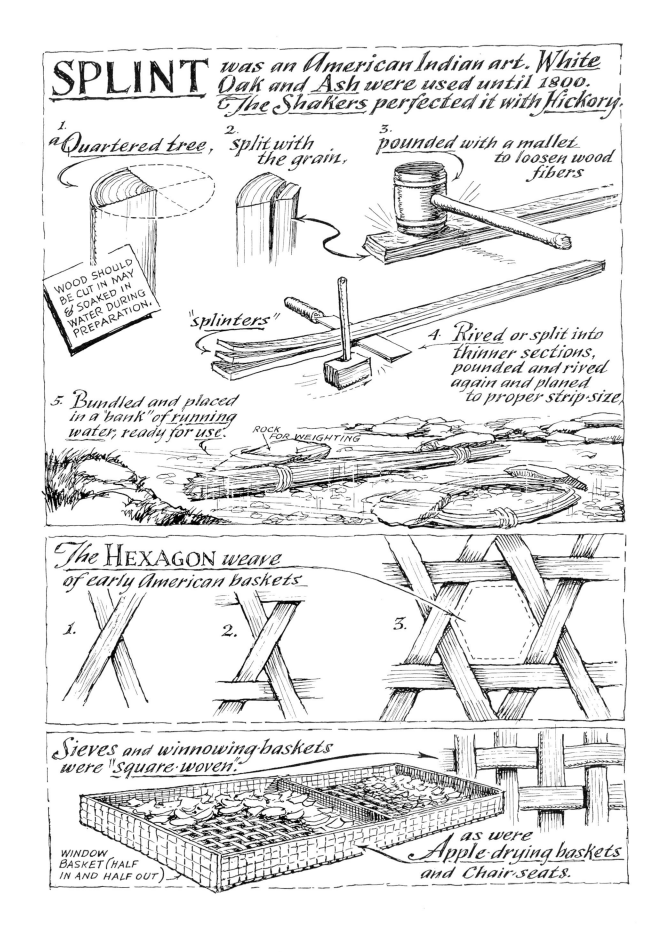

SPLINT

SPLINT was an American Indian art. White Oak and Ash were used until 1800. & The Shakers perfected it with Hickory.

1. a Quartered tree,

2. split with the grain,

3. pounded with a mallet to loosen wood fibers

WOOD SHOULD BE CUT IN MAY & SOAKED IN WATER DURING PREPARATION.

"splinters"

4. Rived or split into thinner sections, pounded and rived again and planed to proper strip-size.

5. Bundled and placed in a "bank" of running water, ready for use.

ROCK FOR WEIGHTING

The HEXAGON weave of early American baskets

1.

2.

3.

Sieves and winnowing baskets were "square-woven."

WINDOW BASKET (HALF IN AND HALF OUT)

as were Apple-drying baskets and Chair-seats.

a slab of split oak, made the best *LATH* material

Stretched out, and Nailed.

edge of woodsmen and farmers, but even of small children. The following is from a sampler composed by a nine-year-old girl in 1816:

> The winter sap resembles me,
> Whose sap lies in the root,
> The Spring draws nigh; as it, so I
> Shall bud, I hope, and shoot.

Another, done in 1780:

> I was born and hoop was cut in May.
> My hoop like me is strong and gay.

Another woodlot chore in May was the gathering of splintwood for baskets and barrel-hoops, and of yellow birch for brooms. The drawing shows how an Indian-type birch broom is made. With a little practice such brooms and shorter brushes can be made in less than one hour, strong enough to last for years. These were the brooms of early America before broom-corn was grown; the messy twig-broom or "witches' broom" was an entirely European design seldom seen in America.

Splint wood and hoop-poles (uncut hoop-wood in pole form) were once a ready cash crop, particularly for the children. Just as children collect returnable soda and beer bottles now, farm children used splint and hoop-poles as their medium of exchange. In May, when black ash and hickory and white oak are porous and vibrant with new sap, six-foot poles were cut from the lowlands and quartered, ready for cutting into splint. These were kept in a "bank" of running water which kept them soft and ready for splitting and pounding into barrel-hoop and basket material. An armful of finished splint was always good for a fair bit of spending money from the town cooper.

Hoop-wood was not merely a part-time farm business. Only fifty years ago there were still large "hoop farms" that raised birch and hickory saplings and cut them in six-foot lengths to be sold at $3.50 a thousand. The size of this business might be indicated by noting that *one hundred million lengths* were sold in Ulster County, New York, in 1898.

Another vanished seasonal industry was the cutting of oak lathing. True, the use of oak lathing was not widespread, but when cut in May and prepared as shown in the drawing the result was a superior building lath that would outlast any other. There is no reason why sheets of composition board could not be cut in the same manner now, and stretched out to any desired proportions. Perhaps no one has thought of it—that is, since the 1700's, when the idea was last used with sheets of oak.

Wood cut in the May season was useless for house framing and finishing, but it had the toughness and rubberlike quality needed for the handles of implements. The "feel" of a hammer or other hand tool is inexplainable, yet no expert craftsman will ridicule the statement that wooden handles are more "comfortable" and "balanced" than metal or plastic ones. Tools were once looked upon as an extension of man's arm, as the worker's genius flowed naturally into whatever he worked upon. A "live" tool-handle added to the craftsman's ability.

June

Delay gives Strength; the tender bladed Grain
Shot up to Stalk can stand the Wind and Rain.
The Tree, whose Branches now are grown too big
For Hand to bend, was set a tender Twig.
When planted, to the slightest Touch would yield,
But now has got Possession of the Field.

ALMANACK FOR JUNE, 1778

If ever a season were chosen to exemplify the virtues of delay, June would be that season. This is the month when the sun is at its zenith, so growing things really take hold. "The slowest plant of June," say the old timers, "will be the biggest soon." The business of spring husbandry has now ended, and with prayers for a slow and strong growing season, summer begins.

June is the season of long daylight; in this month the days grow to fifteen hours long, a long way from December's season of nine-hour days. Summer was once the season of getting things done, unlike the modern season of vacation, which is one of *not* getting things done. Vacation is from the Latin root *vac*—"empty"—which is what good vacations shouldn't be. It is a pity that anyone should waste the longest days in the year doing what they were supposed to do every evening and all day Sunday throughout the year—resting.

Everyone knows that June is the popular season for weddings, but few know the reason why. The custom is actually a superstition connected with the Goddess Juno, who was worshiped by the women of ancient Rome during all critical moments of their lives. A wedding in June is therefore supposed to give mystic protection to the bride. Various places have had different seasons of marriage, according to both fashion and religion. In New York, for example, fashionable weddings were once held between the opening of the Metropolitan Opera season

57

and the first day of Lent; in New Orleans and Galveston, the season was from late November until Mardi Gras.

In throwing rice at a newly-married couple most people think they are wishing them "good luck," but they are really "making them fertile" by showering them with an ancient emblem of fecundity. The rural wedding party as an early American ceremony was an occasion of noise and cider drinking, known as a *shivaree*, at which the guests arrived equipped with noisemakers such as cowbells, horns, and shotguns. The Germans called it a "Callithumpian Band" and the choicest instrument for the occasion was the "Devil's Fiddle" or "Bull Roar." This traditional wedding instrument serenaded the newlyweds with the music of a log-bow drawn over a tightly stretched "string" of hemp rope. The "music" was supposed to be loud enough to announce the wedding to everyone within five-mile earshot. New England Presbyterians followed the Irish custom of firing guns. Their wedding day was suitably welcomed at daybreak at both the bride's and groom's house by a voluminous discharge of musketry, and salutes were fired from every house along the road as the groom proceeded on his way to the church.

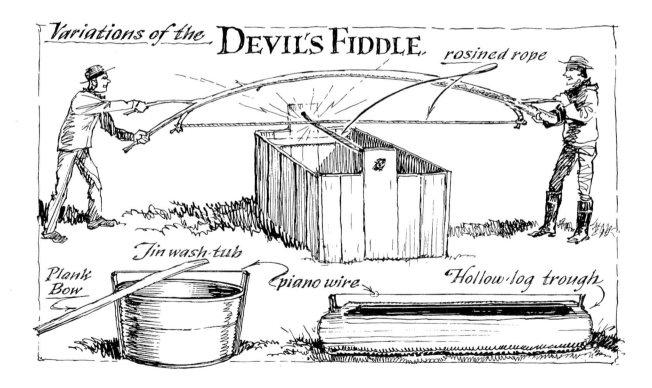

Variations of the DEVIL'S FIDDLE.

rosined rope

Plank Bow

Tin wash-tub

piano wire

Hollow-log trough

If you have an early childhood memory of the leafing countryside, it is most likely a memory of June. To the children went the privilege of finding "forest furniture" and natural implements, those hooks and clothes pegs and other useful items cut from natural tree formations. But farmers themselves often built the strongest and most serviceable things from those natural formations, utilizing nature's ability to "cement" limbs in the tightest fashion to the trunk of the tree. You can devise from natural growths all sorts of benches and racks for hanging things. The drawing shows but a few.

The strawberry season opens the doors of summer. Not long ago, strawberry suppers and strawberry festivals (held during the Strawberry Moon) were always a suitable excuse for summer get-togethers. Strawberries must have seemed much more of a treat when you had to wait for them; perhaps we enjoy them less today for having them available in any season. Wild strawberries were found in America from the start, and as accounts have it, "the Indians bruise them (strawberries) in a mortar, and mix them with meal to make Strawberry bread." A tea made of dried strawberries was one of the first pioneer teas. Strawberry tea, like Sage tea, Catnip tea, Mint tea, Sassafras tea, and Blackberry-root tea, were native delicacies of the past that ushered in the summer "with great medicinal effects." Other popular table teas were Appalachian tea, New Jersey tea, Oswego tea, Crystal tea, and Labrador tea, all once as common as our present-day list of soda-pops.

Sassafras tea is the most familiar on the long list of early American teas. Few people know that sassafras was America's first industry. Ships left our shores with sassafras, at twenty shillings (about four dollars) a pound, as their complete cargo. Odd that the not-too-well-known shade tree, whose leaves and roots children sometimes enjoy chewing, was once known throughout the world as America's prime export.

Another of America's forgotten industries was the manufacture of mint essences. The mention of mint now recalls only juleps or peppermint candy, but a century ago it was the base for most medicines and teas. Spearmint, Watermint, Bergamot mint, Apple-mint, and Pennyroyal were on every housewife's list; but Peppermint was at the top. Still recommended by doctors, it is unrivaled for the alleviation of indigestion. Candied mint (recipe in back of book) is just returning as an expensive specialty, but it was possibly the first American candy.

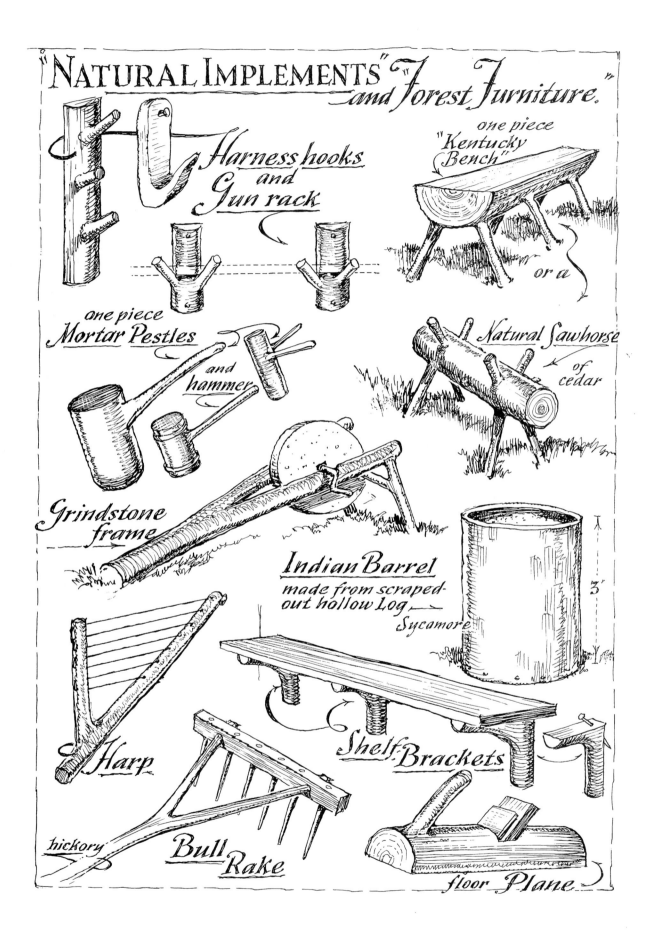

"Natural Implements" and "Forest Furniture."

Harness hooks and Gun rack

one piece "Kentucky Bench"

or a

one piece Mortar Pestles

and hammer

Natural Sawhorse of cedar

Grindstone frame

Indian Barrel made from scraped-out hollow Log — Sycamore

3'

Harp

hickory

Bull Rake

Shelf Brackets

floor Plane

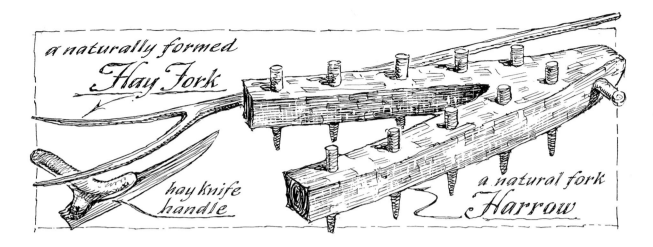

a naturally formed *Hay Fork*

hay knife handle

a natural fork *Harrow*

Most early American summer teas contained pineapple-mint, lemon thyme, or peppermint. A sprig of the herb was sometimes put in the cup and the hot tea poured over it. Mint vinegar (see recipe) was the most common vinegar of the early 1800's.

The well known American mint-julep was originally a mint tea, spiked with liquor for use as a cold remedy. The word "julep" meant "rose water" and the American julep, as picturesque as the old Kentucky Colonel, was at first a morning drink for heavy-drinking Southerners who enjoyed their eye-opener under the guise of "medicinal rose water."

The town of Ashfield, Massachusetts, became the center of America's peppermint industry in about 1813. It was there that the Yankee peddlers went to gather their supply of peppermint, and casks of oil were prepared for shipment to all parts of the country. Peppermint oil sold at sixteen dollars a pound for a while, and many a fortune was made in the industry. But, as in all boom businesses, the bottom soon fell out of the market when the country found itself too well supplied with peppermint. The soil of New York and Ohio was found to be just as well suited to mint-growing as that of Massachusetts, and Ashfield's mint

a Pocket-sized *Mint Mill* of applewood

R.M

Connecticut 1825

an Early List of Basic Herbs from an old Cookbook

Pot Herbs	Vegetable	Flesh	Other	Soup	Fish
Sage	BEANS, ONIONS, TOMATOES.	STEWS, STUFFINGS, SAUSAGE.	EGG DISHES, TEA, TONIC.	SOUPSTOCK, CHOWDERS.	STUFFINGS, BAKED FISH.
Savory	SQUASH, LENTILS, BAKED BEANS, STRING BEANS	VEAL, LAMB, BEEF, STUFFINGS.	EGG-SAUCES.	BEAN, ONION, LENTIL.	CHOWDERS, BAKED, BROILED.
Tarragon	PEAS, BEANS, CELERY, ASPARAGUS, TOMATOES,	CHICKEN, GAME, VEAL, HAM.	ALL EGG-DISHES. MEAT-SAUCE.	VEGETABLE, TOMATO.	SHELLFISH.
Marjoram	ALL VEGETABLES	POT ROASTS, STEWS, CHICKEN, PORK.	CHEESE, GRAVIES, TEA.	SPINACH, MINESTRONE, MEAT.	BROILED FISH CREAMED " CHOWDERS
Thyme	ONIONS, PEAS, CARROTS.	STUFFINGS, GRAVIES, BEEF-STEW.	CHEESE-DISHES, COLD REMEDY.	ONION, CHOWDERS, OYSTER STEW, SOUPSTOCK.	LOBSTER, SHRIMP, OYSTERS.
Bay	ONIONS, SQUASH.	STEWS, LIVER, LAMB.		SOUPSTOCK	PICKLED FISH
Dill	POTATOES, CARROTS, CABBAGE.	LAMB	PICKLES, CHEESE OR EGG DISHES	BEEF, TOMATO	HERRING, SHELLFISH-DISHES.
Basil	EGG PLANT, SQUASH, TURNIPS	MEAT LOAF, LAMB, LIVER, ALL STEWS.	CHEESE EGG OR RICE-DISHES	VEGETABLE, TOMATO.	ALL FISH
Chives	POTATOES.	STEAK, STUFFING.	EGG and CHEESE DISHES	POTATO CREAMED-SOUPS.	ALL FISH
Rosemary	SPINACH, TURNIPS, GREEN BEANS, TURNIPS.	BOILED MEAT, STEWS, VEAL, KIDNEYS.	WITH ORANGE OR GRAPEFRUIT. IN BREADS.	MEAT, PEA, CHICKEN	CREAMED SHELLFISH
Mints	ON PEAS OR CARROTS	LAMB-SAUCES	TEA, VINEGAR, JELLY, SACHETS		
Parsley	USED (BUT SPARINGLY) ON NEARLY ALL COOKED FOODS.				

Anise, peppermint, horehound, angelica for Candies
Savory, costmary, basil, cresses for peppering purposes.
Apple-mint, lemon-thyme, sweet woodruff for fruit drinks.
Coriander-seeds, caraway, sesame, cumin for cake-covering.
Anise, fennel, lovage, chervil, cicely for liquorice flavoring.

"Cut herbs just as the Dew does dry,
Tie them loosely and hang them high.
If you plan to store away,
Stir the leaves a bit each day."

American Farmer, 1842

Seasons of the Common Cooking Herbs.
Choice time to harvest

MINTS *for winter use* . . . . . June and July

SAGE . . . . late Sept., early Oct.

SUMMER-SAVORY, THYME
and MARJORAM . . July and August.

BASIL . . . . August to September

TARRAGON and PARSLEY . . June or July,
or just before flowering.

All herbs should be picked and dried in a Dry, Sunny Season.

HERB DYES of the 1700's

BLUE . . . . Woad, Indigo, Elderberry.

RED . . . . Crimson Pokeweed, Redwood,
Lady's Bedstraw with STANNOUS CHLORIDE.

BROWN . . . . Sumach Bobs, Elder Leaf & ALUM.
Butternut & Goldenrod,
Madder (RED-BROWN), Hop-stalks.

PURPLE . . . . Elderberry with Salt or Alum,
Pokeweed berries, Cedar-tops.

YELLOW . . . . . . Calendula, Cardoon artichoke,
Goldenrod (with stan. chloride), Laurel.

GREEN . . . . Parsley, Elder leaves & ALUM.

BLACK . . . . Gall-berry leaves and Sumach tops.
Scrub-oak & red maple bark.
Yellow-flag (WITH STANNOUS CHLORIDE)

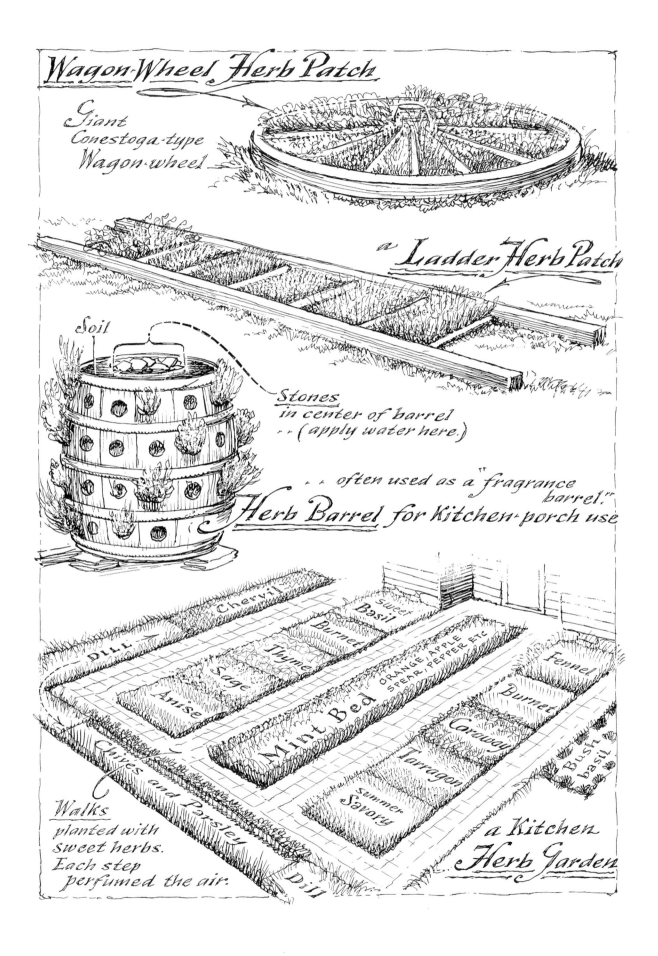

Wagon-Wheel Herb Patch

Giant Conestoga-type Wagon-wheel

a Ladder Herb Patch

Soil

Stones in center of barrel
-- (apply water here.)

... often used as a "fragrance barrel."

Herb Barrel for kitchen-porch use

Chervil

Dill

Sweet Basil

Burnet

Thyme

Sage

Anise

Mint Bed
ORANGE APPLE SPEAR, PEPPER ETC

Fennel

Burnet

Caraway

Tarragon

Summer Savory

Bush basil

Chives and Parsley

Dill

Walks
planted with sweet herbs.
Each step perfumed the air.

a Kitchen Herb Garden

mills soon disappeared. Now peppermint is almost unknown in New England, and the town of Ashfield has probably forgotten this part of its heritage.

Although early America was highly spice-conscious, and no family of taste was without its private herb garden, the art of seasoning foods has frequently become a lost art through the process of haste. American spice chests of a century ago still hold their mouth-watering perfumes, yet most people now seem content with a quick bombardment of catsup or mustard. The seasoning of food has become an almost automatic gesture, rather than the selective art it used to be. You can most times spot an American diner, it is said, by the way he salts and peppers his food before he tastes it.

Not long ago, when nutmegs were welcome gifts and the Connecticut peddler always carried a convenient supply of them in his pockets, silver nutmeg graters were popular; now they are nearly unknown, even to antique dealers. Only specialty shops sell whole nutmegs nowadays, for spices, like everything else, are milled and processed for the mass market. The hanging bunches of herbs that once festooned old-time kitchens are things of the past, but what a wonderful world of smells there must have been! From the cooking of fresh foods and the grinding of rich herbs to the odor of raw aromatic woods and freshly harvested grasses, the person of a century ago had better use for his nose than just breathing.

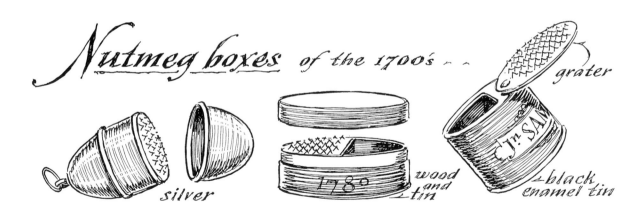

Nutmeg boxes of the 1700's -- grater

silver

1780

wood and tin

black enamel tin

July

Hark! where the sweeping scythe now rips along;
Each sturdy mower emulous and strong
Prostrates the waving treasure at his feet,
But spares the rising clover, short and sweet.

OLD ALMANAC

Independence Day, the birthday of the United States and our season for national rejoicing, was first ushered in by bell-ringing and shooting. Pyrotechnics arrived during the early 1800's, and up until then "fireworks" referred either to the simple art of fire-making or to the Indian's art of shooting flaming arrows. When Chinese firecrackers entered the scene of Independence Day, bell-ringing vanished. But the thought of church bells and farm bells and school bells and fire bells all clanging through the countryside seems best to catch the spirit of that first great American day. It would be fitting, it seems, to revive the bells of Independence Day.

July was known as the season of weeds. More accurately it was the season of weeding, for weeds invade nearly all seasons. Objectionable as weeds can become, their seasonal parade is still a spectacle to fill the real countryman's heart. By virtue of the "forgiveness of nature," where harvests fail or fields are trampled by battle or traffic, weeds seldom fail to cover up the bare ground. Burroughs referred to weeds as nature's makeshift. "She rejoices in grass and grain," he said, "but when these fail, she resorts to weeds."

The lore of weeds is in many ways as interesting as the history of grains, for many plants now considered objectionable were once of great value. In our own time we are seeing honeysuckle, which was one of America's most prized vines, becoming a vicious weed that crowds out

66

all other vegetation that it invades. The troublesome American daisy was first known as the "day's eye." The word was later combined into *days-eye*, and finally to *daisy*. A farm weed now, it was in earlier times prized as first-quality hay material. Dandelion is a bad word in any lawn-maker's vocabulary, yet originally it was an introduced and prized plant; its leaves and buds were used for boiled greens or for "wilted salad," and its roots when ground were used as a good coffee substitute. Summer was "officially in" with dandelion-picking time; but few people of our time have ever tasted dandelions. A beverage may be made from dandelions that was once called Lion-tooth Wine, a name which recalls the plant's original name, *dent de lion,* after the lion-toothed shape of its leaf. Dandelion wine (see recipe) is still considered a fine tonic for elderly people.

The barberry bush, which was so very popular a few years back (and is coming back again) as a landscaping shrub, was not long ago a prime weed. It first came from England, and during Revolutionary times there was rumor that it was sent here to blight American wheat crops; both barberry and wheat, it seems, were prone to the same rust disease. Farmers soon believed that one bush of England's "secret weapon" could blight an entire acre of wheat with rust, and legislation was passed to eliminate the barberry completely from the American countryside. If you saw barberry upon your neighbor's land then, you were permitted by law to enter and remove the bush as a patriotic gesture. This is now a forgotten law, yet it still exists in the statute books of a few states with the provision that "town selectmen may hire persons to destroy any and all barberry bushes."

July has arrived indeed, you realize, when you hear the clatter of lawn-mowers. July and mowing seem to go together, and as if prompted by an urge from the past, man derives some secret satisfaction from the sight and smell of newly cut grass. Haying season isn't an exact date; it's a meeting of weather and man's instinct. The old farmer who never would be guided by modern rules said, "the best time for haying is haying time."

Haymaking by hand was more than just cutting and piling grass—it was an operation that made a man's skill something of rare importance. To begin with, the scythe must be a part of the mower; its weight must match his muscle and its "feel" must be conducive to a graceful swing. The result is a rare poetry of motion that cuts the most grass with the least exertion. The fascination of feeling the sharp blade slice through grass and lay it gently down is pleasing to the observer and satisfying to the mower. By holding the snath-heel low to the ground and the blade delicately tilted upward, windrows of scented grain can be laid in parallel rows with admirable precision. The sight of a group of men "cradling" a field of grass was as exciting as watching a rowing crew, and it all but called for applause. Unlike most work that tires one set of muscles and causes weariness, when a man is tired from scything he is pleasantly fatigued all over; when he rests, he rests all over.

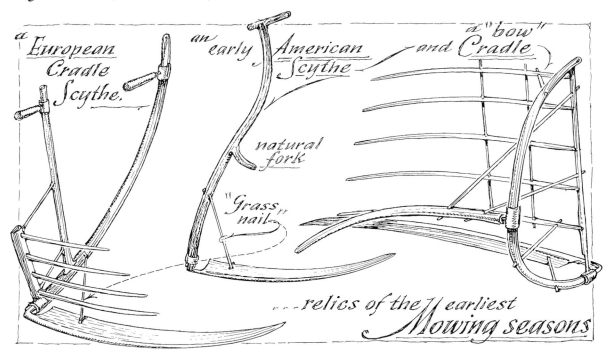

a European Cradle Scythe. an "early American Scythe a "bow" and Cradle natural fork "Grass nail" ...relics of the earliest Mowing seasons

Mowers never left the field for the midday meal, but practiced the old custom of "nooning." A nooning dinner was always a stout one, suitable for a hard-working man, and there was usually a jug of switchel to wash it down with. Switchel was a molasses-based sweet drink that belonged to the early harvest season. It could be "spiked" with cider or brandy, but Haymaker's Switchel was usually made plain in a three-quart keg for taking afield. To two quarts of water were added: 1 cup of brown sugar, a half cup of sweet vinegar and about a teaspoonful of ginger. If the keg leaked ever so slightly, the evaporation only helped to cool the drink.

The farmer's nooning added to the richness of the old-time harvest day, and it lasted at least a full hour. Farmers respected their midday

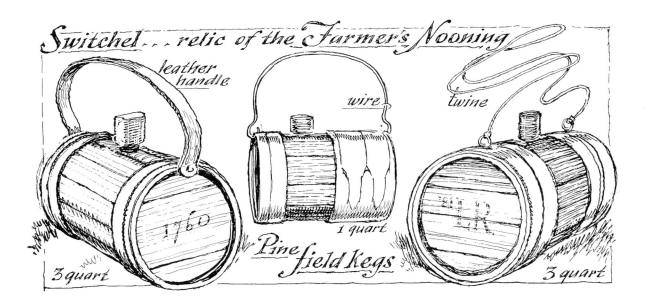

Switchel...relic of the Farmer's Nooning

leather handle

wire

twine

3 quart

1760

1 quart

Pine field kegs

3 quart

mealtime, and there was always a shock of hay at hand for relaxing in. Lowell's "Indian Summer Reverie" recalls this scene:

> The wide-ranked mowers wading
> to the knee—
> Then, stretched beneath a rick's
> shade in a ring,
> Their *nooning* take.

Thomas Tusser mentioned the barley cradle in 1560, but that "short-fingered" design was quite different from the graceful "long-fingered" cradle which was an American device. Few people now appreciate cradle-scythes as Americana, and antique shops usually destroy them rather than allow them to take up so much space. But, like the plow, their lines are a symphony of pure utilitarian grace. No other implement so well expressed its maker or called for greater skill in the making; the cradle was undoubtedly the Stradivarius of early American farm implements.

The four or five slender ash root "fingers" of a fine cradle were chosen carefully from the forest for their matching curve, and seasoned by charring and scraping during the winter months around the hearthside. The snath or handle was selected from a properly twisted root or crooked tree that must have been hard to come by. After the 1860's handles were made of willow, shaped in hot oil, and dried to the desired form. Altogether, a good cradle might be the result of many months of labor, but to the farmer it was always worth the effort. When one examines one of

69

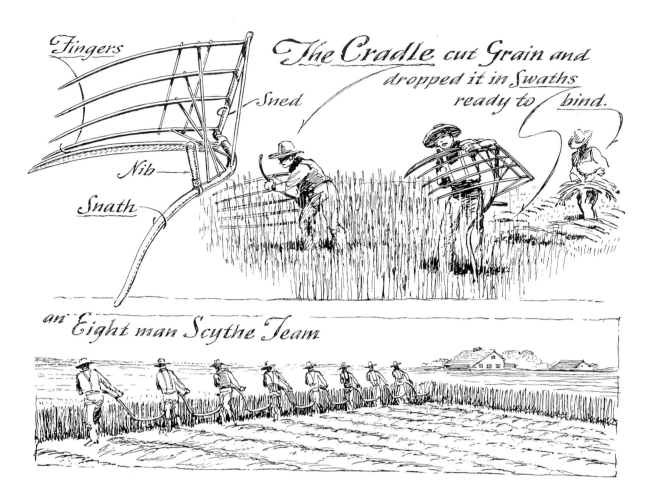

Fingers

The Cradle cut Grain and dropped it in Swaths ready to bind.

Sned

Nib

Snath

an *Eight man Scythe Team*

these antique cradles, it seems almost a miracle that an implement of such delicate curves and webwork could have cut and swathed hundreds of tons of grass, and yet survived two centuries.

Another vanishing piece of farm Americana is the circular stack. Most people today refer to simple corn shocks as "stacks" but a corn shock is merely a group of field stalks tied into a standing bunch. A stack, however, was an orderly and architecturally built pile made so as to turn

The Stack Roof was adjustable

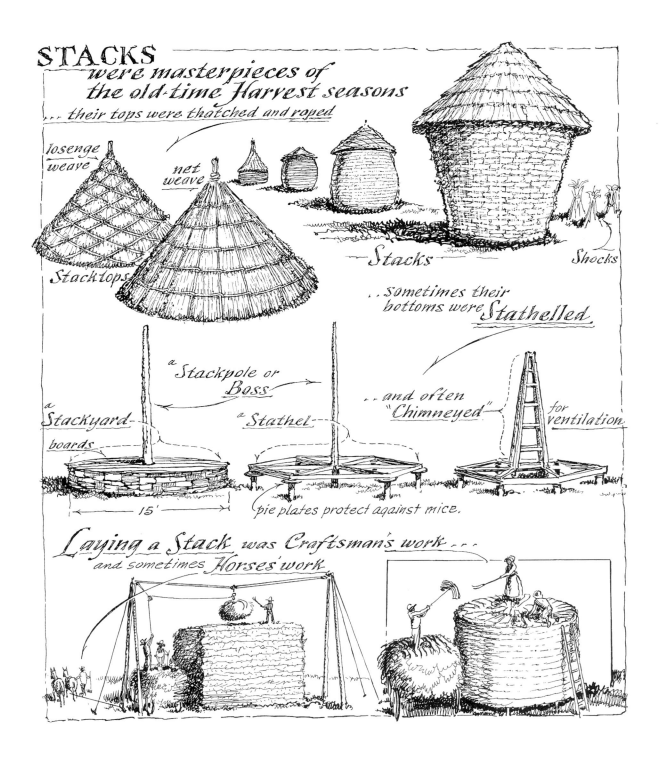

STACKS
were masterpieces of the old-time Harvest seasons
… their tops were thatched and roped

losenge weave

net weave

Stacktops

Stacks

Shocks

… sometimes their bottoms were *Stathelled*

a *Stackpole or Boss*

a *Stackyard boards*

a *Stathel*

… and often "Chimneyed"

for ventilation

15'

pie plates protect against mice.

Laying a Stack was Craftsman's work …
and sometimes Horses' work

rainwater outward and downward, and covered with a thatched roof. Every farmer had his own design, and the subtleties of stack building were once as distinctive as those of any other rural architecture.

71

Marking the high noon of the year, watermelon time was a July season of great importance a few years back. It was a very brief midsummer season, for in those days watermelons were religiously left on the vine until ripe and then eaten at once. The difference between a vine-ripened watermelon and one ripened en route to the market would surprise most people, as the flavor is greatly impaired by early picking. Properly ripened, a melon shrivels and blackens its tendril nearest the fruit, then yellows its underside. When flicked with the finger it resounds with a characteristically dull thud.

Negroes were supposed to be singularly fond of watermelons, a myth that started because American watermelons grew only in Florida until about a century ago, and also because they originated in Africa. There, they were often used as a "water supply," grown to ripen in drought time and collected in large quantities as an emergency measure. An endless series of nineteenth century "minstrel" jokes would have had you believe that the American Negro just couldn't resist watermelons. It *was* the Negro, however, who first utilized castaway watermelon rinds to make one of America's choicest preserves, *watermelon pickle* (recipe in back of book). "Watermelon cakes" were a July specialty in the 1800's. They were not made of watermelon, but of cake batter tinted red to resemble it, and raisins were used to resemble the seeds (see recipe).

A separate book could be written on America's seasonal cooking and vanished delicacies. Only the hot-cross bun seems to remain, for pumpkin pies, mince pies, and many one-time seasonal breads are now baked all year round. "Hot-cross buns" were first known just as *cross-buns* and they were eaten only on Good Friday, the cross, of course, symbolizing Christ's crucifixion. Commencement cake (see recipe) was a fruit-and-wine cake, eaten on commencement day at the colleges and often made in the shape of a small book. Election Day cake (see recipe) was a raised fruit cake baked only on election days; it originated in Hartford, Connecticut, in 1839 "to be served to all those who voted the straight ticket," and it was sold in New England bake-shops for over half a century.

The old custom of baking a cake for a journey has changed with the modes of transportation, but "journey cakes" (a term corrupted later into *johnny cakes*) were the particular pleasure of small children (and still might be revived as a more desirable traveling companion than gooey candy bars and ice-cream on a stick). Journey cakes were also baked en

a few of the old time **SEASONAL CAKES**

red coloring and raisins

Journey board

baking a *Journey Cake*

Watermelon Cake (July)

fire

rock

Easter Cake

Election Cake

Election Day

Straight Ticket

and *Hot cross bun*

Commencement cakes "like little books"

crust

Christmas cake was the original mince-pie baked "open-face" in a dish fashioned after the Christ Child's manger. The chopped meat and fruits represented gifts of the Wise Men.

Manger Dish

* holly was once an Easter emblem

route by travelers on horseback, who carried their own "journey-board," a scoured wooden shingle with a handle on it. Used like a baking tin, it *slanted* against the camp fire. Johnny cake is best known as a simple cornmeal mixture, but there was also a "journey cake" so-called because it needed neither milk nor eggs (which were seldom transported), but used cider as fluid. (See "cider cake" recipe.)

Simnel was named after Simnel Sunday or "Mothering Day," which is the fourth Sunday of Lent. This was the holiday for apprentices and housegirls, who often traveled great distances to visit their mothers. The Simnel cake (see recipe) was their traditional "calling card." It is presumed that Anna Jarvis hadn't heard of Simnel Day when in 1907 she suggested the second Sunday in May as American Mother's Day. What was the florist's gain was the baker's loss, for mother is out her Simnel cake. The old English version of Simnel cake was topped with candied violets, which attests to Britain's milder weather and earlier spring.

Simnel... cake of the first Mothers' Day. candied violets — paper wrapping.

It is interesting to note that in the old days cakes were often packed in popcorn for transporting. If they were crushed there was no loss, for the icing-smeared popcorn could still be eaten. Here is an idea that might still be used today!

A teacher was explaining to her young class what the "staff of life" is. "What is it," she began, "that American children eat every day?" "Candy," was the answer. When America was young, however, the present form of candy was unknown. Candy then was a verb meaning "to crystallize or break into bits," and *sugar candy* was no more than broken sugar. A large piece of sugar was served in a dish, along with a pair of "sugar-scissors" or a "candy hammer," neither of which can be found in antique shops today.

Sugar cakes or *marchpane* (marzipan) was a seasonal treat, although in many respects it was the forerunner of our present-day candy. The list of

Sugar scissors and Candy hammer

seasonal marchpane designs seems almost endless. The sugar, egg-white, and nut-meat recipe was usually the same, but the cakes were shaped to fit the occasion. There were cakes for Christmas, Epiphany, Easter Eve, Ash Wednesday, Maundy Thursday, and about sixty other holy days, while the patriotic and vocational designs ran into the hundreds.

The carved wooden mold blocks for making marchpane were mostly made in Europe, where the idea originated, but those few molds that are made of native American walnut will some day be among the rarest and most valuable Americana. Bakers and confectioners (or "sugar-bakers," as they were first called) had an average of fifty to a hundred designs in each shop; whatever became of more than a quarter of a million molds remains a mystery.

a Marchpane Mold for Independence Day

mold-maker's mark

Baker's Initials

August

See where the farmer, with a master's eye,
Surveys his little kingdom and exults
In sov'reign independence. . . .
Here stacks of hay, there pyramids of corn,
Promise the future market large supplies:
While with an eye of triumph he surveys
His piles of wood, and laughs at winter's frown.

1753, ROBERT DODSLEY

The early English farming season began in January, on Plow Monday. It ended on the first day of August, which was called Lammis Day, so it became the logical Old World Thanksgiving day. Being the farmer's thanksgiving, Lammis Day persisted in early America with all the enthusiasm of rural holiday spirit, until a succession of American National Thanksgiving Days caught up with it. In 1863, when Lincoln proclaimed a National Thanksgiving Day, Lammis Day vanished forever from the American calendar.

Lammis Day was a logical and meaningful holiday which, it seems, should not have disappeared, because the first harvest of the year seems so ideal a time for thanksgiving and celebration. On this first day of August, the old-time farming family always went to church. The head of the

LAMMIS DAY
the Lammis Loaf!
the Lammis sheaf
August first

76

household dressed in his best, and brought with him the first loaf of new-grain bread for consecration, that loaf later becoming the center of a feast not unlike our present-day Thanksgiving dinner. Sometimes it was the first sheaf of wheat harvested or the first pack of corn that was blessed and feted on now-forgotten Lammis Day.

August is often the season of drought in America, so it was once the season for digging wells and planning water systems and, of course, for *water witching*. Where there was water in August, the countryman knew there would be water all year round. "Divining," "dowsing," or "water witching" was never considered a superstition among the old-timers; it was regarded as an inexplicable yet entirely scientific process. "The same power that leads a willow's roots to water yards away," they insisted, "will also pull a willow fork if held by a man who really believes.

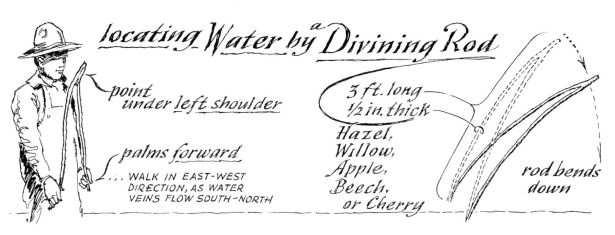

locating Water by a Divining Rod

point under *left* shoulder

palms forward

... WALK IN EAST-WEST DIRECTION, AS WATER VEINS FLOW SOUTH-NORTH

3 ft. long — ½ in. thick

Hazel, Willow, Apple, Beech, or Cherry

rod bends down

A good dowser can even lay his hand on a disbeliever and make a divining rod work for him, too." Divining rods were used for locating almost anything beneath the surface of the ground. Lost coins, for example, were hunted by placing a coin in a slit at the fork of the rod. "It takes silver to find silver," the saying goes.

A fork of hazel, cut in drought time, made the most sensitive of all witching-rods. Swamp alder and swamp willow were also used, but the medicinal and mystic qualities of American hazel, plus the legend that magicians of ancient times had used hazel wands, made witch hazel the standard wood for dowsing.

Trees having a limb bent peculiarly downward and then upward, it was observed, usually had a vein of water running beneath the limb. The

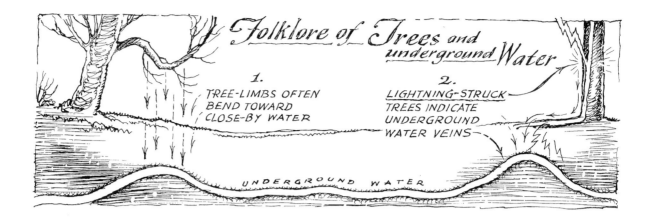

Folklore of Trees and underground Water

1.
TREE-LIMBS OFTEN
BEND TOWARD
CLOSE-BY WATER

2.
LIGHTNING-STRUCK
TREES INDICATE
UNDERGROUND
WATER VEINS

UNDERGROUND WATER

explanation was that when the tree was a sapling the waters "witched the limb downward," but as the tree gained in strength the limb resisted the force and began to grow upward again.

Another piece of water lore was that a tree struck by lightning is supposed to mark a vein of water; this has been found to have scientific support, as lightning experts now agree. Early farmers, therefore, chose not to build their house or barn over a vein of water (for fear that the building would be "water grounded" and more attractive to lightning), and actually used a dowser for locating where water *wasn't!*

One of the vanished pieces of farmyard equipment is the well-sweep. By way of restoration, or just for decorative purposes, people frequently try to build these without a knowledge of the true mechanics and proportions of a real well-sweep. The object of a well-sweep was to let the sweep lift a full bucket by its own counterweight; the bucket, therefore, had to be small and the sweep itself enormous. The average sweep was either a long, heavy pole or, preferably, a whole length of tree, often over forty feet long; the bucket usually held two gallons or less. Instead of a rope, a *drop-pole* held the bucket, and with it you "pulled the bucket" into the well by lowering the drop-pole hand over hand.

No water supply, it seems, was really as satisfactory as the gravity spring. It never froze in winter because it ran constantly; for the same reason it was always cool in summer. The giant hogshead that it emptied into in the old-time kitchen held a constant and inviting supply of fresh, cold water. In these days, when the country house water supply stops if

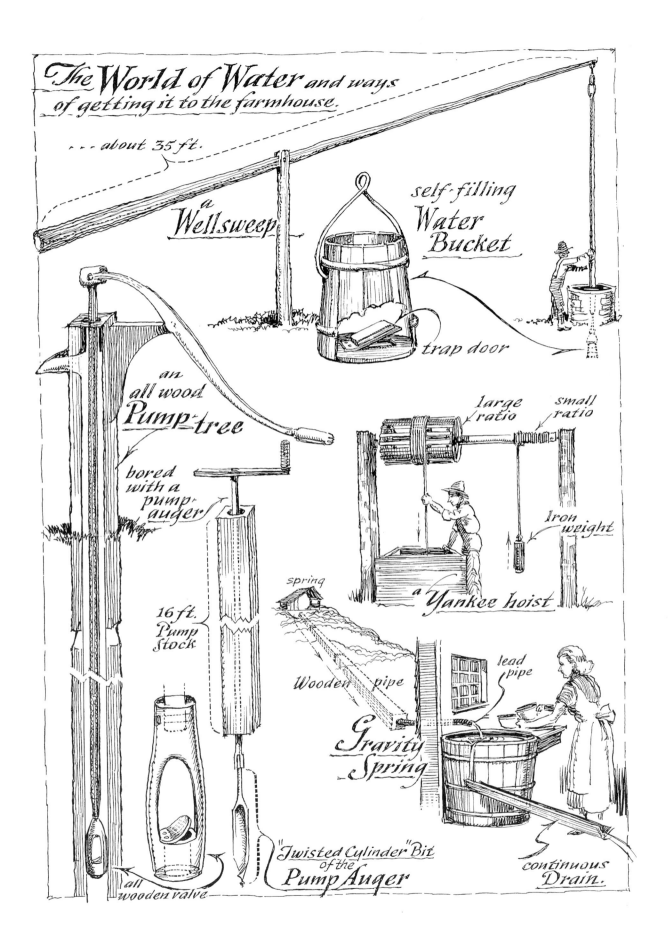

The World of Water and ways of getting it to the farmhouse.

--- about 35 ft.

a Wellsweep

self-filling Water Bucket

trap door

an all wood Pump-tree

bored with a pump-auger

large ratio   small ratio

Iron weight

16 ft. Pump Stock

spring

Wooden pipe

a Yankee hoist

lead pipe

Gravity Spring

"Twisted Cylinder" Bit of the Pump Auger

all wooden valve

continuous Drain.

the electricity goes off, the "obsolete" gravity spring system shows its superiority over the modern system.

Whereas well-sweeps were for outdoor open wells, the old "attached wells" were divided into two classes, *porch-wells* and *breezeway-wells;* both were under cover and handy to either winter or summer kitchens. In the picture of the porch-well, notice that the underground room for keeping things cold (cellar) was situated under the summer kitchen and opened by "windows" into the cool well-pit. The convenience of running water, of course, is here to stay, but open wells and cool cellars have a certain nostalgia about them that man will always find delight in.

a *Fence-post Vise* and a *Fence-rail Vise*
made from a crotch and a stick —

Fence-post Drill

Fence Post Axes (1800's)

For driving rail-holes into fence-posts

When a historian said that "Americans discovered drinking water," he wasn't being as funny as he may have thought. In the 1600's, the Dutch drank beer at the table and the English drank ale; the Spanish and French drank wine. No one dreamed of drinking water at meals in those days because to drink water in the New World was considered a dangerous experiment. The earliest pioneers used molasses-beers and fruit ciders at the table, but by 1800 water had been "discovered" by the farmers, and it then found its way to the American dinner table.

The month of August, it seems, was once one much devoted to charity. One might argue that charity should enjoy a year-round season, as it usually does today, yet even now we have only one thanksgiving day and many other single occasions for godliness and good will that might better be expressed every day of the year. Perhaps it was the overflow of harvest crops that prompted such activities as August "bees" and gatherings to raise money or collect provisions for the poor, but the Bible prompted the season, too. "And when ye reap the harvest of your land," said the Lord, "thou shalt not wholly reap the corners of thy field, neither shalt thou gather the gleanings of thy harvest . . . thou shalt leave them for the poor and stranger." (Lev. 19:9, 10) So was it an early farming custom to leave a small open corner of the field nearest the road, for the poor.

The famous Millet painting "The Gleaners" appears to be a scene of women harvesting and the word *glean* has generally been misinterpreted as a synonym for *harvest*. Gleaning was an ancient custom of the poor, who picked up the scraps of harvest, and *gleaners* were therefore the needy. The sign of the zodiac called the *Virgin* was often represented not as a reaper but a *gleaner*, and the season at that time of August was one devoted to charities "as divined by the stars."

In Europe, the season of gleaning was widely accepted, even regulated by law. The custom was still practiced in the early days of America. A church bell rang or a drum was beat, usually in the last week of August, signaling the time when the needy could move toward the unfenced fields and gather what they would of all harvest scraps.

We might not now understand why Europeans in the 1700's called us "a nation of savage fence-builders." But realizing that fences were at that time classified as military devices (from the word *defence*), and that a fenced field seemed designed to exclude even the poor gleaners, we can better imagine how the American farmer with his heavily fenced farm might have appeared to the European farmer.

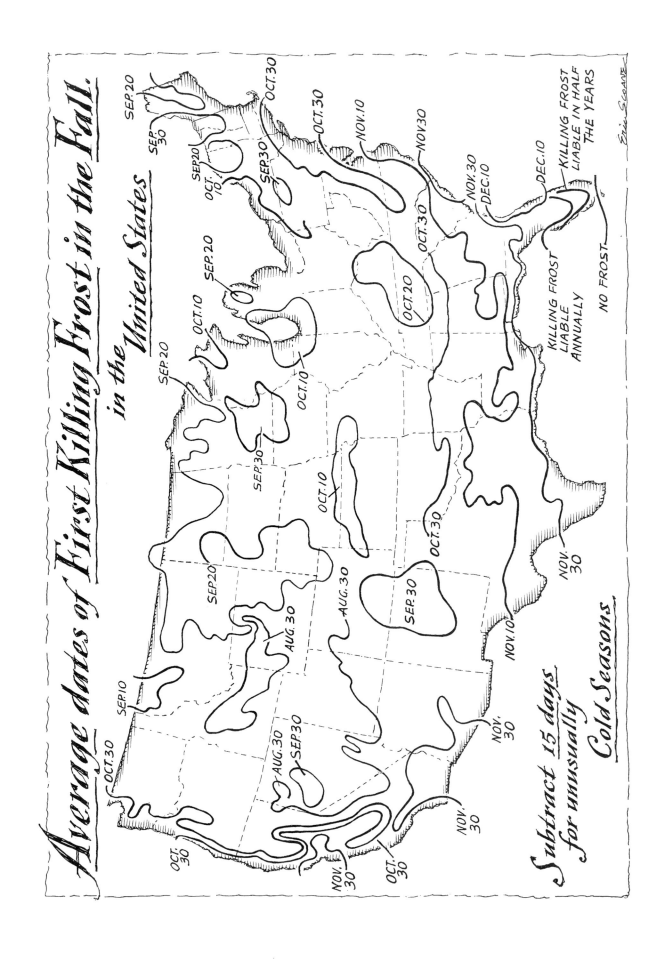

Average dates of First Killing Frost in the Fall.
in the United States

Subtract 15 days
for unusually
Cold Seasons

KILLING FROST
LIABLE
ANNUALLY

KILLING FROST
LIABLE IN HALF
THE YEARS

NO FROST

Erin Sloane

September

Poets call autumn the melancholy season, but to American farmers it was the season of fulfilment and a time of rejoicing. Why else would they have chosen September as the Season of Fairs? The only melancholy of September is experienced by school children, who realize better than anyone else that vacation is done; or perhaps, if houses have souls and are capable of melancholy, by those closed-up summer places that look lonelier than lonely in their autumnal surroundings.

The most typically American of all seasons might well be Indian Summer. A phenomenon of the autumn season, with no fixed date, it ranges from September through November. It is a mystic warm spell that occurs after the first frost of "Squaw Winter" and before the entrance of actual winter. It has been called Late Summer, Sham Summer, the Fifth Season, All Hallow's Summer, Red Man's Summer, and Smoke Season. It is a brief season of quiet beauty that occurs when most people are "back from vacation," and the fewest Americans, therefore, are at liberty to appreciate it.

Legend has it that Indian Summer was so named because of the atmospheric haze present at that time, and that the pioneers associated it with *Indian war fires*. Actually this was the season when the Red Man went into the interior to prepare for winter hunting, and the oft mentioned "Indian fires" were only those used for scaring game into traps and groups of hunters. The blue haze of Indian Summer is caused by salt

83

particles within the air that settle during the autumnal change of high-altitude prevailing wind patterns.

This season of haze, the last sweet smile of the declining year, still reigns from coast to coast as the most American of American seasons; its magnitude has far superseded its historical interest through Indian lore. This was the farmer's brief season for a vacation from work, for relaxation by hunting, and fishing, and exploring. And what better time exists than this, when nature and weather combine to put on the greatest show in America?

Hunting seasons are nothing new, but the early farmer's reverence for nature nearly always kept him from killing more than he needed to kill. Of the tons of equipment found on the old farm, the sparse equipment for hunting and fishing is worthy of note. A Damascus shotgun or a flintlock rifle converted to cap and a set of bullet molds often constituted the American farmer's complete arsenal. Those farmers whose religious beliefs ruled out guns had none at all.

Trapping was boy's sport on the farm, and to capture animals as the Indians did called for far more skill and gave greater personal satisfaction than gun-killing can offer. The sling-shots, arrows, and snares which used to delight all country boys have lost their appeal now, but they were part of a popular fantasy of self-preservation that children enjoyed in the days when forests and game were close by. Even adults played at it, for nearly every decade had its pioneer who entered the

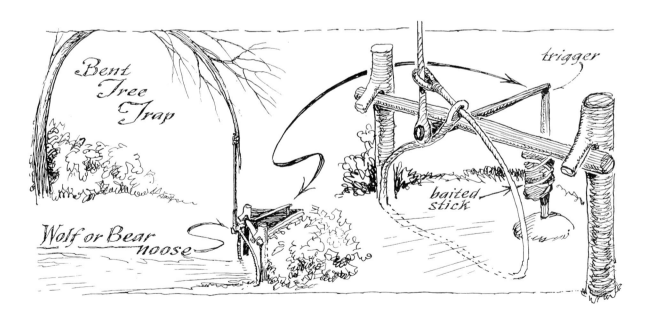

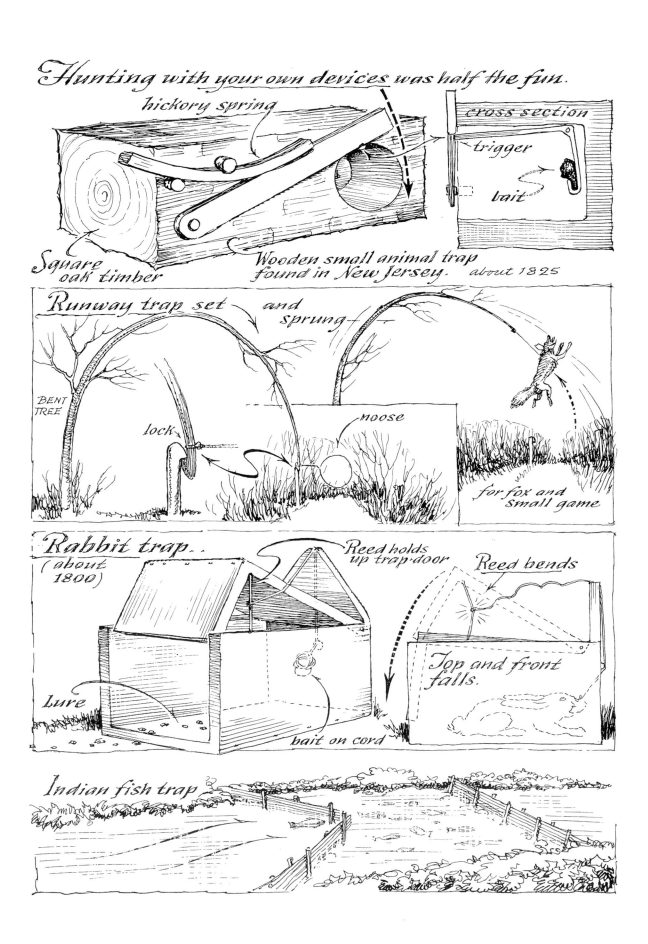

Hunting with your own devices was half the fun.

hickory spring

cross section

trigger

bait

Square oak timber

Wooden small animal trap found in *New Jersey*. about 1825

Runway trap set and sprung

BENT TREE

lock

noose

for fox and small game

Rabbit trap.
(about 1800)

Reed holds up trap-door

Reed bends

Top and front falls.

Lure

bait on cord

Indian fish trap

forest naked and returned some time later well fed and fully clothed. It was a fad that lasted for over a century, and small boys listened breathlessly to tales of their adventurous heroes, learning how to recognize edible berries and to build snares for birds and animals. One of the more famous romanticists who felt the challenge of the wilderness was John Ledyard, who left Dartmouth College in 1772 to try his luck at primitive survival. Clothed only in a bearskin, he made a wondrous voyage down the Connecticut River in a hollowed-pine canoe. Children still find this type of appeal in such stories as *Swiss Family Robinson*, but matches or cigarette lighters make a fire quicker than rubbing two sticks together, and the hard way of doing things has finally lost its challenge.

The first game law in America called for a closed season on deer in Massachusetts in 1694, and in 1739 the first "deer wardens" were appointed. In 1818, Massachusetts prohibited the killing of robins in the spring of the year, when robin-pie had become a national delicacy. New York was the first state, in 1864, to adopt a hunting-license law. But because many Americans could not restrain the temptation to make everything possible an article of commerce, the "balance of nature" soon ended for much of our wildlife. Foremost is the story of the passenger pigeon, which, as we read, seems beyond belief. Here is an account of the September of 1826:

> The banks of the Ohio were crowded with men and children, for here the pigeons were to fly low. For a week or more the population spoke of nothing but pigeons and fed on no other flesh but that of pigeons. The whole atmosphere was strongly impregnated with the smell appertaining to their species.

> Allowing two pigeons to the square yard, a column one mile in breadth will often pass overhead for over three hours: the count in one such flock should be one billion, one hundred and thirty-six thousand pigeons. . . . The dung of such a flock becomes several inches thick, covering the extent of their roosting place like a bed of snow. Many trees of two feet in diameter have been observed broken by their weight, at no great distance from the ground.

> The sun was gradually lost to view, though not a pigeon had yet arrived. But all of a sudden I heard a cry of 'here they come!'

The noise which they made, though distant, reminded me of a hard gale at sea. . . . As the birds arrived, I felt a current of air that surprised me. . . . Thousands were soon knocked down by poles. . . . The crashing of trees breaking from roosting pigeons added to the din.

from: Times Telescope, 1828

The last passenger pigeon died in the Cincinnati zoo on September first, 1914.

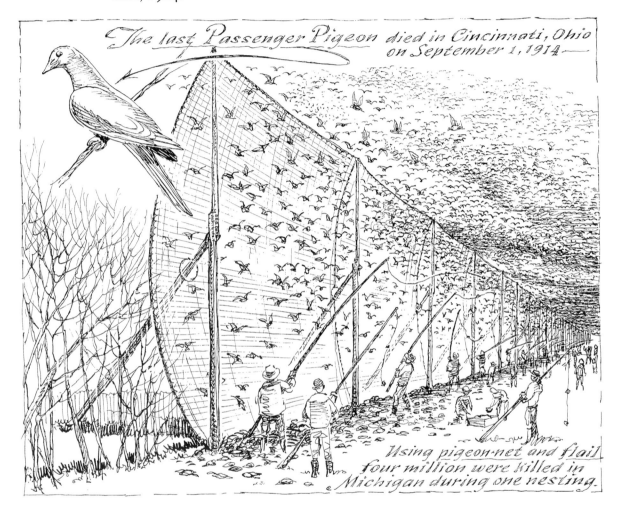

The last Passenger Pigeon died in Cincinnati, Ohio on September 1, 1914—

Using pigeon-net and flail four million were killed in Michigan during one nesting.

One of America's most popular game birds, the Chinese pheasant, was first brought here by none other than George Washington. In 1789 he imported several pairs from England for his Mount Vernon estate and tried to tame them—unsuccessfully, for the pheasant is shy and untamable. Today these birds are found in all the states, multiplying just a little faster than they are killed by hunters.

Fishing was one of the prime country contentments and pleasures of old-time autumn, particularly for young boys. Pins for hooks and horse-

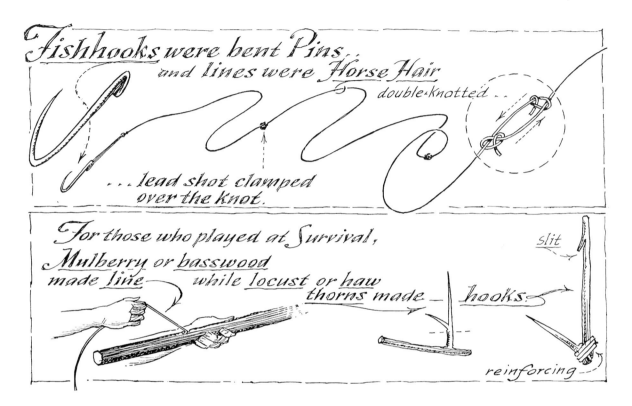

Fishhooks were bent Pins...
and lines were Horse Hair
double-knotted...
...lead shot clamped over the knot.

For those who played at Survival,
Mulberry or basswood
made line — while locust or haw
thorns made — hooks
slit
reinforcing

hair for lines were sufficient, for fish were most often jerked from the water rather than "played." Gray or white stallion's hair was considered the best line, and only that longest "haire which groweth from the middle and in most part of his dock" was chosen. The line was often weighted with lead shot clamped over each horse-hair knot. If you could have the shot "bitten onto the line by a fair maid," the legend went, your fish line was blessed with luck and you were assured of good fishing.

"In the Fall of 1807," wrote Elkanah Watson, "I procured the first pair of Merino sheep that had appeared in Berkshire. . . . I was induced to notify an exhibition under the great elm tree in the public square at Pittsfield (Massachusetts) of these two sheep on a certain day. The farmers present responded . . . we became acquainted, and from that day to the present, agricultural societies, cattle shows, and all in connexion therewith have predominated my mind." That small meeting of farmers and two sheep was probably the first American Fair. From it grew the Berkshire Agricultural Society and the institution of American fairs. European fairs, and those of early colonial days, involved the sale of cattle and produce, but the Great American Fair became a farmer's holiday and general exhibition where prizes were given for products and handiwork of the community.

The downfall of the old-time American agricultural fair to the level of the modern American carnival is regarded by many as a national disgrace which belittles the institution of farming. The introduction of a circus midway has made the agricultural exhibits only a side show to the fair, and many of America's oldest fairs have degenerated into collections of barkers, salesmen, freak shows, Ferris wheels, and "girly shows." The farmer's carnival of today is far removed from the institution it was a century ago.

In many latitudes where the early apple season falls in September, picking apples was the children's chore just before school's opening. It might be frowned upon as child labor nowadays, yet it might just as well be recommended as adolescent therapy. It was always a delight to the children and a blessing to the farmer. "Let your children gather your apples," said the first American *Farmer's Manual.* "[Children] are the farmer's richest blessing, and when trained to habits of industry, they become the best members of society, when they grow into life." Amen to that! "Let them eat apples, too," it continued, "for nothing will strengthen and preserve young teeth more." In 1957 the National Apple Institute recommended apple eating as an aid to tooth-cleaning. "With an apple," they said, "you eat your toothbrush."!

Hoops were one of the early spring toys, but along about the middle of September children played hoops again. Like marbles, which suddenly appeared on an unannounced day, hoops reappeared, as if by some strange signal, in late September. The answer is probably that in the early days a large part of the day consisted of the long walk to and from school, and hoops could be rolled along the way. A hoop was a companion to a child just as a cane can be to a man. But hooping was an art, too.

89

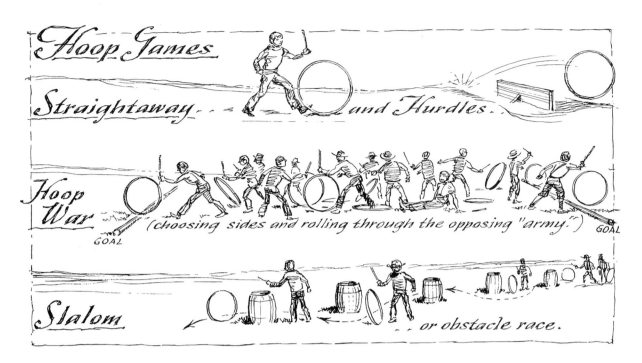

Hoop Games

Straightaway ... *and Hurdles*.

Hoop War (choosing sides and rolling through the opposing "army.") GOAL GOAL

Slalom ... or obstacle race.

There were hoop "straightaway races," and hoop "slalom races," and hoop "wars." In a hoop war, two groups of boys set against each other and tried to drive their hoops past the "enemy." All hoops knocked down, of course, were counted as "casualties."

The back-to-school season automatically became the time for organizing new tug-of-war teams. Tug-of-war was a game taken more seriously (even with grown-ups) than we might now imagine. Every village and fire brigade and police department and merchant's association had its tug-of-war team, and to be a member of such a team made you one of the more respected citizens. Children merely tried to pull their opposing team across a chalk-line or through a ditch of water, but adult teams who performed at town fairs and picnics were drawn through everything from pits of gooey mud to burning ashes. Women entered tug-of-war tournaments too, and strangely enough they often won against the men,

Heave ho! *There goes the ladies' team!*

not because they were stronger but simply because they were heavier. Just as tall boys are chosen for basketball nowadays, the town fat folk were eyed as tug-of-war candidates.

The pickling season adds final zest to all the savory smells of September. Today the word *pickle* brings to mind a prepared cucumber, but *pickle* in the old days was a verb that referred to the brine or process and not to the actual product. You pickled beets, watermelon rind, tomatoes, corn, and an endless variety of fruits and vegetables.

There are four basic types of cucumber pickles: *dill, sour, sweet,* and *quick. Dill pickles* are produced by putting freshly picked whole cucumbers in spiced vinegar. *Sweet pickles* are sour pickles drained of vinegar and aged in a sweet spicy liquor. *Quick* or *"overnight pickles"* have neither brine-curing nor fermentation (see pickle recipes).

Today we "put up" preserves, but they used to be "put down" into the cellar. One of our more misused words has been *cellar*, which meant

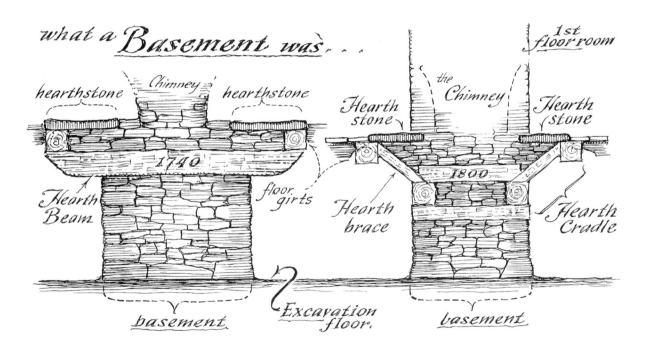

what a Basement was . . .

hearthstone   Chimney   hearthstone   the Chimney   1st floor room

Hearth stone   Hearth stone

1740   1800

Hearth Beam   floor girts   Hearth brace   Hearth Cradle

basement   Excavation floor   basement

food-locker until the advent of the icebox. Cellars were underground rooms (not usually beneath the house) where the temperature ranged from 38° in winter to 50° in summer. To keep dampness out, the cellar often had its own chimney, and the dry richness of the old-time cellar, perfumed by smoke ovens and harvests of the past year, must have been quite different from the stale mustiness of cellars as we know them.

We have tidied up the cellar, and now call it a "basement." The true meaning of the word *basement*, however, has long been forgotten. A *basement* was once the *foundation* for the chimney; the "basement room" was that hollowed-out portion beneath the house where the "basement" stood. The drawing shows two typical "basements" with their supports for first-floor fireplaces. When you realize that many early houses were built around (and were almost held up by) the center chimney, you can see how important basements were in the old sense, and why a whole "room" was often given to them.

It was once believed that putting food below the level of the ground helped to preserve it. Generally speaking, there is truth in the theory. Silos were one time not the lofty round buildings of today's farm, but holes in the ground (as evidenced by the word *silo* itself, which is derived from the French word for "pit"). People also believed that the ground insulated things against the "harmful effects of moonlight." Pickling and canning, for example, were never done during moonlight. This belief is not as ancient as it sounds, for here is a hard-to-believe report from the U.S. Weather Bureau (Department of Agriculture) in *1903:*

> That moonbeams or rays produce certain chemical results seem certain. It is known that fish and some kinds of meat are injured or spoiled when exposed to the light of the moon.

Picking tomatoes by the light of the moon was once supposed not only to make them more palatable, but to make the eater amorous. When at first tomatoes were called "love apples," the Puritans outlawed growing or eating them as sinful. Their culture was attended by a long spell of bad luck before they became accepted. Thomas Jefferson started growing tomatoes in 1781, but in trying to popularize them he unfortunately mentioned that they are of the nightshade family (some species of which are a poison), and many people thus considered tomatoes poisonous.

The old-time method of leaving tomatoes on the vine to ripen is not practiced commercially, so millions of Americans have never tasted vine-ripened tomatoes. Tomatoes are usually picked pink today, and often given a shot of ethylene gas to give them proper color. But they lack flavor, and their vitamin content is sometimes reduced by as much as thirty percent.

September is the season of the Harvest Moon. The full moon that falls nearest the autumnal equinox (on or about September 21) is in that part of its orbit where it makes the smallest angle with the horizon. For several nights in succession the moon rises at nearly the same hour, giving an unusual proportion of moonlight nights. Since it rises slower, the "huge" effect of the moon is exaggerated, and the harvest moon is therefore supposed to appear larger and redder than the moon of any other season. Many a harvest has been worked in the open field only by the light of a full September moon. The full moon of October, next in light scope, is known as the Hunter's Moon. "The moon of September," says an early almanac, "shortens the night. The moon of October, is hunter's delight."

October

Now o'er his corn, the sturdy farmer looks,
And swells with satisfaction, to behold
The plenteous harvest that repaid his toil.

ALLEN'S ALMANACK, 1821

One week after the first October frost, the sorghum season began. A poor-soil brother of the corn family, sorghum grows all over the United States and as far north as Canada. To midwestern and southern mountain folk, in the days when they knew sugar only in liquid form, there just wasn't any other sweetening like it. Sorghum meant a rich dark-brown molasses, just right for corn bread and unbeatable for hot-cakes. It is still used for seasoning beans and for making cookies. Very much like the northerners' maple-sugaring time, a sorghum "run-off" was the most enjoyable farm event of the old-time farm year. Sorghum was mostly a small-farm product, but during the Civil War years about sixty million gallons of it were manufactured. Today sorghum has been bred into a dry-soil plant for livestock feeding, and yesterday's small-farm "sugar-plant" has become a main crop of the Middle West. Although Kansas is still known as America's "wheat state," it now grows more sorghum than it does wheat!

The beers mentioned in early American writings were in no way similar to beer as we know it—and such was southern molasses beer, made from sorghum. A first distillation of fermented sorghum juice, molasses beer was found on the tables of most mountain farms, often as a substitute for milk, and was taken by small children at every meal.

The typical Kentucky family had two acres planted in sorghum. Most of it was for syrup, part went for cattle fodder, and the seeds fed the chickens. The sheet metal pan for cooking the syrup was similar to New

England's maple-sugar pan, but the horse-drawn sugar mill originated in the South. Northerners usually preferred to do their "farm squeezing" with wooden screw-type presses. Squeezed sorghum juice exuded from the mill through a burlap strainer and into a barrel. It was then transferred to the cooking pan. As the juice began to boil, it was paddled and cleared of impurities, turning from green to muddy and finally to clear brown. Four gallons of juice produced about one gallon of syrup; as a substitute for store-bought sugar, sorghum was an easily grown crop with very little waste. Unlike today's sugar with its nutrients refined away, primitive sorghum syrup was not as good to look at, but it at least contained food value. Sorghum joined corn as one of the staffs of early farm life; it even found its way into paints and dyes. (See recipe for sorghum whitewash.)

Cane sugar was introduced into the deep South just before the 1800's, and because of its very long growing season plus slave labor, sugar soon outdid tobacco, becoming America's great new industry. It was the first sped-up farm work in which the measured seasons of nature disappeared

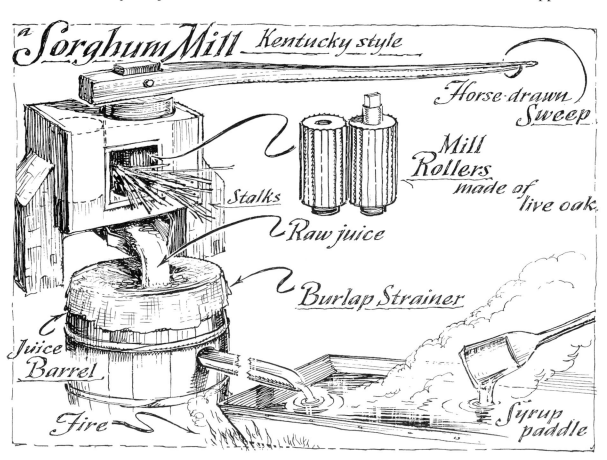

a Sorghum Mill Kentucky style

Horse-drawn Sweep

Mill Rollers made of live oak

Stalks

Raw juice

Burlap Strainer

Juice Barrel

Fire

Syrup paddle

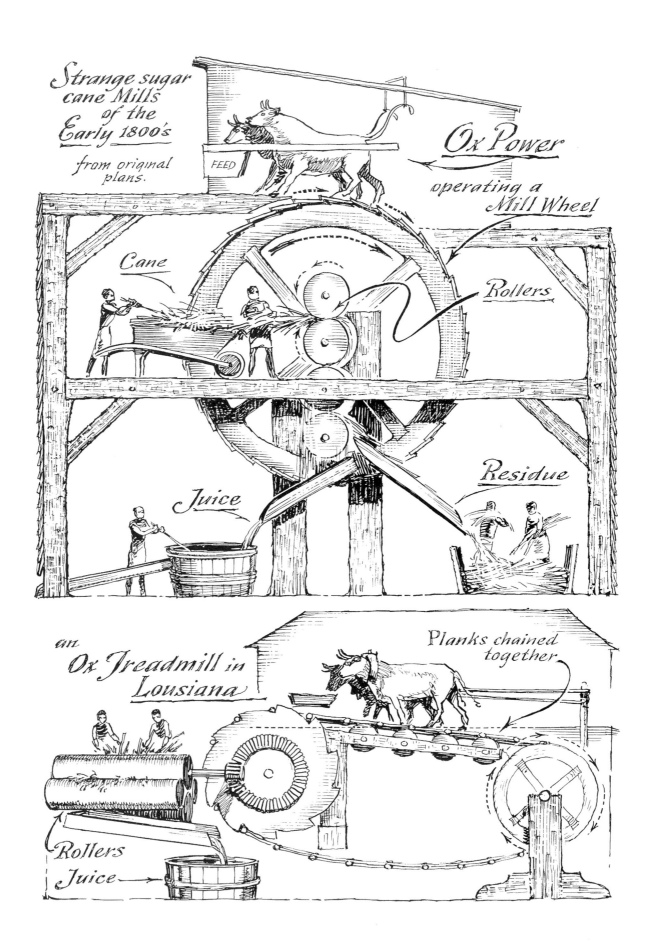

Strange sugar cane Mills of the Early 1800's

from original plans.

FEED

Ox Power

operating a Mill Wheel

Cane

Rollers

Juice

Residue

an Ox Treadmill in Lousiana

Planks chained together

Rollers

Juice

and machinery invaded the farm. The extent to which the first cane-sugar farmers went to mechanize their farms ranged from amusing to amazing. Lacking water power such as the North had, the South ran its machines by groups of slaves, horses on treadmills, and even oxen on the top of a millwheel. The list of cane-sugar mill contrivances is endless, but two examples are shown in the drawing. Such mills have long since rotted and disappeared. Only the ruins of ancient horizontal rollers remain, dotting the deep South countryside like monuments to the industry they began.

As naturally as sunshine follows rain did seasons progress on the farm. When winter apples were all packed into underground mounds with dry sand or hay, stored in cellars, or sliced and dried in splint driers, the proper season for cider and apple butter had arrived. Those apple-butter crocks that nearly filled the early American farm shelf have for no good reason nearly vanished. Cider is in many ways on the way out; although it was a year-round drink up to fifty years ago, it has already become a seasonal novelty. Few of us, if any, have heard of perry (cider made of pears) or peachy (cider made of peaches), yet at one time, both were favorite American drinks. They are discarded now, simply because of the time and effort involved in their preparation. Peachy was known as

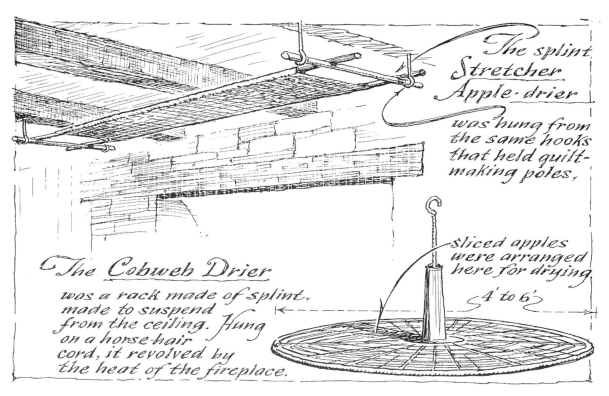

The splint Stretcher Apple-drier was hung from the same hooks that held quilt-making poles,

sliced apples were arranged here for drying

4' to 6'

The Cobweb Drier was a rack made of splint, made to suspend from the ceiling. Hung on a horse-hair cord, it revolved by the heat of the fireplace.

97

the American champagne. It was the ultimate in fine-tasting drinks, to the extent that one described anything very superior by calling it "peachy." What a wealth of taste and goodness we have been deprived of in this day of artificially flavored and sweetened soda waters!

Apple cider or butter is now very often the product of apples that would not sell as eating apples. On the old-time farm, such apples were given to the cattle who relished them, bruises, worms and all, and only the very best apples went into cider and apple butter.

Few farm products are more confusing to the outsider than cider; its definition and recipe vary with the locality. Technically, cider is a fermentation of apple juice containing from one half of one percent to eight percent alcohol. American "cider mills," therefore, can legally sell only *apple juice*, and both the juice and fermentation are loosely known as "cider." Pasteurization stops the fermentation process nowadays, and it also keeps the grocers' shelves from blowing up. But connoisseurs of the great American drink we no longer are.

It is strange that in America, where cider was only recently a national drink, most people now believe cider was made simply by squeezing whole apples. The most important step in cider-making was the *milling* that occurred before pressing. In the mill, apples were chopped and bruised into a rich pomace called "cheese." This seems at first like an unnecessary step, because the press appears to do about the same thing shortly afterward. But let an ancient farm book explain it:

> If the juice of an apple be extracted without first bruising the fruit, it will be found thin and defective in richness, compared to the juice of the same apple after it has been exposed to the air and to sunlight in a bruised state. It then becomes deeply tinged, less fluid and very rich. In its former state it apparently contained very little sugar; in the latter a great quantity. Even by bruising the apple more slowly, a difference in quality is again noticeable.

Cider machinery, then, consisted of two machines—a mill and a press; never was one machine missing. Both were usually of breathtaking proportions, with hewn timbers larger than are seen anywhere today. In Pennsylvania and the north central states there have been beam presses with two-foot-square oak timbers, the giant wooden screws masterpieces

Apple Cider was a man-sized job—
first came the Milling—

a Knob Mill

SOCKETS
hickory rollers
KNOBS

Pomace for "Cheese"

horse-drawn Sweep

fresh apples

a "Grindstone" Mill for apples

a Double "Grindstone" Mill.

Pomace

apple Shovel

an Apple Barrow

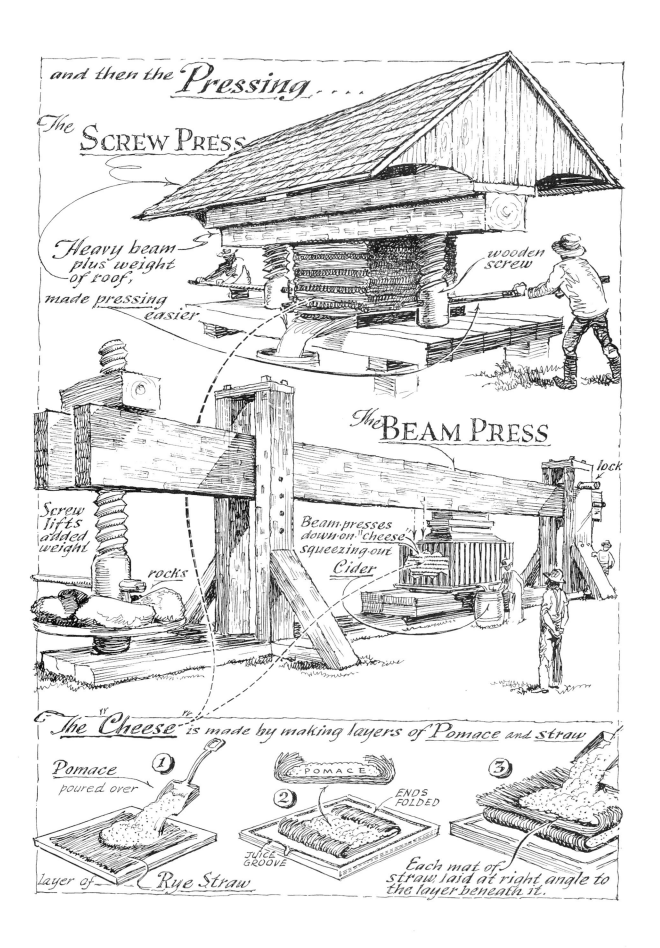

and then the *Pressing . . . .*

The SCREW PRESS

Heavy beam plus weight of roof, made pressing easier

wooden screw

The BEAM PRESS

Screw lifts added weight

rocks

Beam presses down on "cheese" squeezing out Cider

lock

The "Cheese" is made by making layers of Pomace and straw

Pomace poured over

①

② *POMACE*

ENDS FOLDED

JUICE GROOVE

③

layer of Rye Straw

Each mat of straw laid at right angle to the layer beneath it.

of heavy wooden craftsmanship. Even after metal hardware was plentiful, apple machinery remained entirely wooden, because any metal at all was supposed to contaminate cider and spoil its flavor. Intricate apple mills and presses of many tons were made which have lasted for centuries, without as much as one piece of metal, even a nail.

Most antique shops have giant iron kettles such as were used for scalding water at butchering time. These kettles were also used during the apple-butter season. The fact that apple butter had to be boiled and stirred for about eight hours made it an all-day family affair, with hard work and laughs for everyone from grandmother to the youngest child. Along about the sixth hour, when the butter began getting dark, a batch of biscuits went into the nearest oven and everyone got set for sampling time. The chilly end of day, the smell of hickory fires, warm apple butter on hot biscuits, and a well-earned appetite made for a memory that wasn't easily erased; it was the noon of a rich American season. Apple-butter bees are still held in the Middle West; in New England, where they really originated, however, they have now completely vanished. (See recipe for apple butter.)

Apple butter Paddles

The October Butter Season

Johnny Appleseed's favorite apple, the "Rambo," originated in Pennsylvania in about 1800. It was yellowish-white, striped with red, and exceedingly rich in flavor. Still grown on private farms in the Appalachian Belt, it has now nearly gone from Ohio, where Johnny had once started it growing on every farm. The family apple tree, like the family estate, seems to be going out of fashion in America; apple trees, like houses, change owners frequently or succumb to neglect in the changing pattern of the American countryside. A sad shame it is, for no tree lends more to the landscape or yields so much fruit.

Anyone can appreciate the loss of certain food values by the food-packaging methods of recent years, but few of us can believe that fruit itself has changed. The truth is that the best flavored and most nutritious fruit is only that which ripens on the tree, and the average child of the twentieth century has yet to eat such! If today's farmer picked his fruit ripe from the tree, it would not reach the nearest town market before becoming soggy and bruised, so in order to ship fruit hundreds or thousands of miles, as we do now, it must be picked long before nature has properly prepared it to be eaten. The fruit must then do the best job of ripening that it can, en route or on the fruitstand. The loss? Small, yet definite, and more important to health than the sped-up fruit industry acknowledges to its public. This knowledge is nothing new. Thomas Tusser in the 1500's said:

Fruit gathred too timely wil taste of the wood
Will shrink and be bitter, and seldome proove good.

The fruit-grower of a century ago grew fruit entirely for aroma, succulence, and general goodness. Today's fruit-grower cultivates his trees to bear heavy crops of fruit that looks attractive on the stand and transports well; these requisites, unfortunately, often exclude the very best varieties. The Golden Russet, the Tolman Sweet, the Snow Apple, and some two hundred other apples are among examples of "outmoded fruit." Those finest of peaches, the Crawford, Yellow St. John, and Oriole are others on the list. Currants and gooseberries can still be found in country markets, but because they are worth less than the man-hours they take to pick, their market, too, diminishes each year. Even so far as fruits alone are concerned, how rich were the seasons of yesterday!

November

At Hallowtide, slaughter time entereth in,*
And then doth the husbandman's feasting begin.
THOMAS TUSSER

Thanksgiving, November's gayest season, had obscure beginnings. The best known "first Thanksgiving" of 1621 was not a single day, as traditionally pictured, but a whole week of gaming, rejoicing, and feasting. Another "first Thanksgiving" was proclaimed in Charleston, South Carolina, and June 29 was designated "the day for thanking God for deliverance." One famous Thanksgiving Day was February 22, 1630, in Massachusetts, a *fast* day! Famine, it seems, was imminent, so a certain day had been set aside for fasting and praying for deliverance. But at a critical moment ships with provisions appeared, and the fast day was quickly proclaimed a day of thanksgiving. Another interesting early American Thanksgiving Day came in the spring, when people gave thanks for having lasted through the cold and deprivations of winter.

The first proclamation of Thanksgiving for the United States was made in 1789 by President Washington, who set aside November 26 as the proper day. In 1863 Lincoln made it the last Thursday in November, and Franklin Roosevelt made it the Thursday before that. All these dates, which are still presumed to be times for giving thanks for a bountiful harvest, are well past the opening of harvest seasons. The pumpkin has become an American Thanksgiving emblem largely because it is one of the few things surviving above ground on the farm at so late a date!

---

* All Saints' Day, November first.

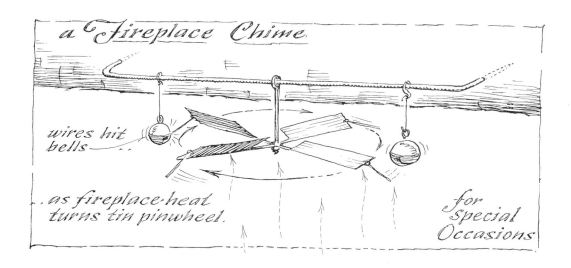

a Fireplace Chime.

wires hit bells—

...as fireplace-heat turns tin pinwheel.

for Special Occasions

The end of November was for many early farmers a special season of feasting. It ended the general droving season, and cattle money was usually plentiful. The droving season opened on June first and closed in November, but considering the old roads, which were most often rivers of mud, the dry autumn months were the best time, and often the only possible time, for driving cattle to market. Some of the last big November droves must have been rare sights. Farmers often stopped work and left the fields to watch the parade. A cloud of dust and a thunder of animal noises heralded the approach of a driven herd as much as five miles away. There might be a thousand each of hogs, sheep, and cattle, with turkeys, geese, or goats between these three sections. Their average speed was between seven and nine miles a day; a drove might take a month or more to reach market. Although still practiced in 1900, droving's golden age was about 1830. In 1850 it began to decline, with the projecting of canals and railroads, and that great Knight of the Road, the Drover, began his slow disappearance from the American scene.

K-o-o-o K-o-o Boss!

104

The head drover usually led a "lead-ox" and gave the traditional cry of "Ko-o-o Boss! Ko-o-o-o Boss!" Although our word *boss* originally came from the Dutch word *baas*, for master, "Boss, the lead-ox" of the old-time drove is said to have contributed most to our present Americanism for "the head man," and the drover himself was often referred to as "boss." The term "Bossy, the cow" also stems from the old droving days.

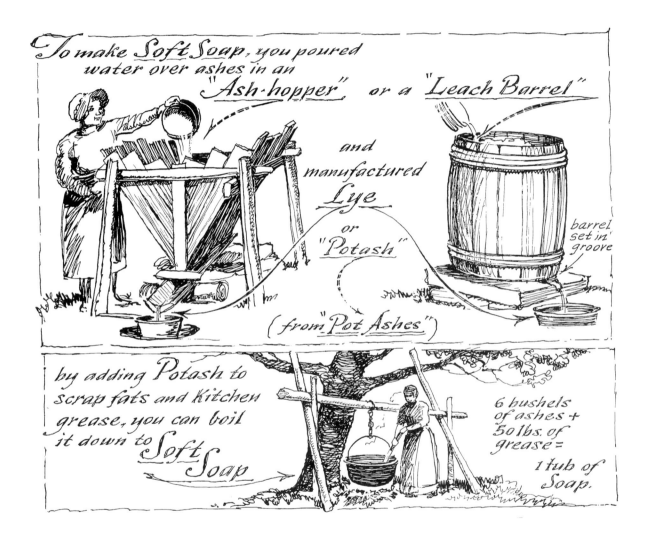

To make Soft Soap, you poured water over ashes in an "Ash-hopper," or a "Leach Barrel" and manufactured Lye or "Potash" (from "Pot Ashes")

barrel set in' groove

by adding Potash to scrap fats and Kitchen grease, you can boil it down to Soft Soap

6 bushels of ashes + 50 lbs. of grease = 1 tub of Soap.

Early spring and late fall on the farm were generally soap-making seasons, but as soap was made mostly from grease and fat, November's butchering-time made the fall season the more popular one. On small farms where no butchering was done, table scraps and old lard were saved all winter and used for soap-making in the spring. As repulsive as these ingredients sound, they produced the clear jelly that made soft soap much more desirable than one might think.

Soft soap was made from lye and grease, and every family had its own "ash hopper" or "leach barrel" for manufacturing lye. This was a sticks-and-straw-bottomed container for wood ashes. Only certain hard woods made proper ashes: pine ashes, for example, were considered worthless. With four quarts of wet lime tamped into each barrel of ashes, water (greasy dish-water was fine) "leached" through the ashes and came out the bottom as lye. If you are a back-country pioneer who has suddenly run out of store soap and are now carefully following this recipe, be sure

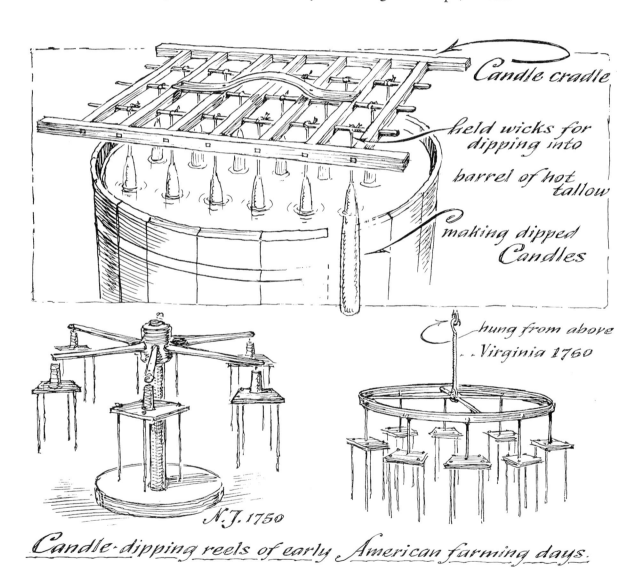

Candle cradle

held wicks for dipping into

barrel of hot tallow

making dipped Candles

hung from above
Virginia 1760

N.J. 1750

Candle-dipping reels of early American farming days.

to get your lye strong by pouring it, if necessary, over fresh ashes at frequent intervals. If an egg or potato floats in it, the lye is just right. Now you must boil your lye and grease together in a big pot and presto! soap!

But, "before the soap becomes a mass," you might remember, "scent it well with sassafras." And give it a ten-week seasoning period before using it. Running out of soap used to mean a long time between baths.

The soft soap of yesterday, along with its wooden container, has long gone. Now you can't even find soap containers in antique shops. But soft soap is beginning to appear again in tins and plastic containers, and, of course, with lanolin and other modern "magic" ingredients.

The candle-making season was in November (or early December in the South). It had to be just cold enough for quick hardening, and followed close after "killing-time"; it was usually waste fat that made candle tallow. Tallow candles burned brighter than today's wax version but unless properly prepared they produced a black smoke. They were most often a blend of ox and sheep fat, cut into pieces and melted or "rendered." The fat was boiled, caked, pressed, sieved, and purified several times. Wicks were made of cotton, but a rarer and finer wick was spun from milkweed.

Tusser listed candle-making as a November pursuit, and advised:

Wife make thine owne candle,
Spare pennie to handle.
Provide for thy tallow, ere frost cometh in,
And make thine owne candle, ere winter begin.

We are most familiar with the early metal candle molds, such as are now used as lamp bases, because they were so well made and lasted so long, but candles on the farm were more often dipped than molded. Thousands of wooden candle cradles and other candle-dipping equipment are destroyed yearly just because no one knows what they are. The drawing shows a candle cradle with the wicks tied to it; the wicks were dipped again and again into a tub of hot tallow, and each time, of course, the candles became bigger.

Bayberry candles were made during late autumn, when the berries were ripest. The bayberries were thrown into a pot of boiling water, and their fat rose to the top and became a superior candle wax. Bayberry

Bayberries, boiled in water floated their wax to the surface. Skimmed and cooled in moulds they made a very special Candle.

a Wax-myrtle scoop

candles burned slowly; they didn't bend or melt during summer heat, and yielded a fine incense, particularly when the candle was snuffed. So prized were bayberry candles that the gathering of berries before autumn in America once brought a fifteen-shilling fine.

The silver-gray berries of scented bayberry, known in England as the "tallow shrub," were for many years sent overseas as Christmas souvenirs from the New World. In the 1700's, the bayberry was more Christmasy than holly (which represents the thorns and blood of the *crucifixion* rather than the *birth* of Christ). The burning of a bayberry candle at Christmas was as traditional in America as the burning of a yule log in England. "A bayberry candle burned to the socket," an old verse goes, "brings luck to the house and gold to the pocket." Children seldom went to bed on Christmas night without the magic charm of a bayberry candle, and the perfume of the snuffed bayberry candle was part of that magic night.

We might forsake all the old ways and live in an electrified world of chrome-plated conveniences, yet if we want the ultimate in graciousness we still reach into the past and light candles.

One of America's saddest seasonal losses is that of chestnut time. In late October and early November, boys and squirrels were rivals in the gathering of autumn's treasure trove; chestnutting time arrived when the spiny chestnut burr burst open. Roasted chestnuts were delicacies in themselves, yet chestnuts were also used with meats, in cakes or bread, and even as a base for pudding (see recipe for chestnut pudding).

As if nature knew the preciousness of chestnut meat, the chestnut is clothed in a silk wrapper, enclosed in a case of sole leather, packed in shock-absorbing, vermin-proof pulp, sealed in a waterproof iron-wood box, and finally encased in a covering of porcupine spines. "There is no nut so well protected," said Ernest Thompson Seton, "and there is no nut in our woods to compare with its food."

When a native chestnut blight began from a Japanese exhibition in the New York Botanical Gardens in 1904, it resulted in one of our greatest losses. It killed every tree as it spread. The old houses and barns built entirely from chestnut are still monuments to the time when chestnut rated with pine and oak as useful American woods. There are groups of farmers who remember the old chestnut days, who are trying to revive the American chestnut tree. To date there are reports of several chestnut trees thirty feet tall.

An American November isn't complete without cranberries. They were discovered by the Pilgrims in America, where the Indians were using them for food, as a medicine, and as a dye. Almost at once they were recognized as a preventative against scurvy, and the early sailors kept a good supply of cranberries in all ship stores. Now we know that it was their vitamin C content that was responsible for the berries' good effects.

Cranberry harvest with Berry Scoops in Massachusetts

1830   Cranberry Farmers' Turf Axes   1810

The origin of the word *cranberry* has been strangely lost, but as cranes once lived in the bogs and ate the berries, there is some reason to believe that they might have first been called *craneberries*. They were also called bounce-berries, because their ripeness was tested (and still is) by their ability to bounce.

In 1816, Henry Hall of Dennis, Massachusetts, observing that cranberries grew biggest wherever sand blew into their beds, went about building his own sanded cranberry farm. His experiments culminated in the first real cranberry bog nearly fifty years later, which, incidentally, still produces today. The so-called New England industry has now spread down-coast and out to the Pacific. Oregon, Washington, Wisconsin, and New Jersey are all centers of the cranberry industry, but two thirds of the nation's crop is still produced on Cape Cod.

December

The piercing cold commands us shut the door,
And rouse the cheerful hearth; for at the heels
Of dark November, comes with arrowy scourge
The tyrannous December.

ALLEN'S ALMANACK, 1821

Stephens' "Book of the Farm" (1840) says, "Winter is the especial season of man—*our own season*. It is the intellectual season during which the spirit of man enables him most to triumphantly display his superiority over the beasts each day that perish." In winter, the countryman plays a conqueror who sets forth each day to battle the elements and, winning, returns to the rewards of his harvests. It's a daily game, beyond the ken of the city-dweller whose comings and goings lack the flavor of make-believe.

The coming of winter in the old days was heralded by a "banking-up season," when the north sides of houses and barns were stacked with proper insulation. Corn stalks, hay, leaves, or sawdust shouldered the base of the farmhouse against winter's blast; cow dung did the job best at the barn. Up north it was hemlock boughs. Near the seashore it was seaweed. A coast-farm December chore was to gather windrifts of seaweed and pitchfork it to a height of about two feet, enough to cover the sills and underpinning of the house. Where the wind did creep in, it carried with it the salt tang of summer days. But without the "bank-up," few of the early farmhouses could have survived the cold of northern American winters.

The first of December was Sled Day wherever winter and snow were synonymous; that was the day when sleds of all sorts were readied and sleigh bells were made to shine. Just over sixty years ago there were real sounds to winter: steel-shod runners squeaked over the packed snow and

111

Winter Banking
Seaweed or dry leaves
Corn-stalks
16" pine Bank-board
about 1 ft.

the almost constant music of sleigh bells filled the crisp air everywhere. Winter was a season of bells.

Time was when you could recognize a neighbor's approach by the sound of his sleigh bells, even tell which neighbor it was. Some farmers made up their own sets of bells and others preferred to use inherited sets. For those who wished to buy, however, there were Swiss Pole Chimes, Mikado Chimes, and King Henry Bells; the Dexter Body Strap of twenty-four bells was a popular buy. At first, sleigh bells were made from two half-globes of metal soldered together, but one-piece bells were later cast and sold separately, ready for fastening to harness. A matched set cost about $1.50.

East Hampton, Connecticut, once a parish of the town of Chatham, was known as Jingletown, the sleigh-bell capital of America. Here America's first globe-type sleigh bells were made, and were distributed throughout the world. They were sold by the pound.

The reason for using bells on a sleigh was not only for merriment but primarily for safety. A sleigh was a silent vehicle and a fast one, which its driver often found the greatest difficulty in stopping. Furthermore, everyone wore ear muffs or some other sort of ear-covering in the early days, so that winter pedestrians were practically deaf. Just as lights and horns are now required on the highway, bells were once a "must" for all winter traffic.

Although we are all familiar with the lore of early American stage-coach days, we seldom hear of *stage sleds* now. Used they were, however, and the designs of stage sleds were particularly interesting. They were often insulated, padded with velvet, curtained, and heated by charcoal burners. Most of them were even equipped with toilet facilities. Stage-

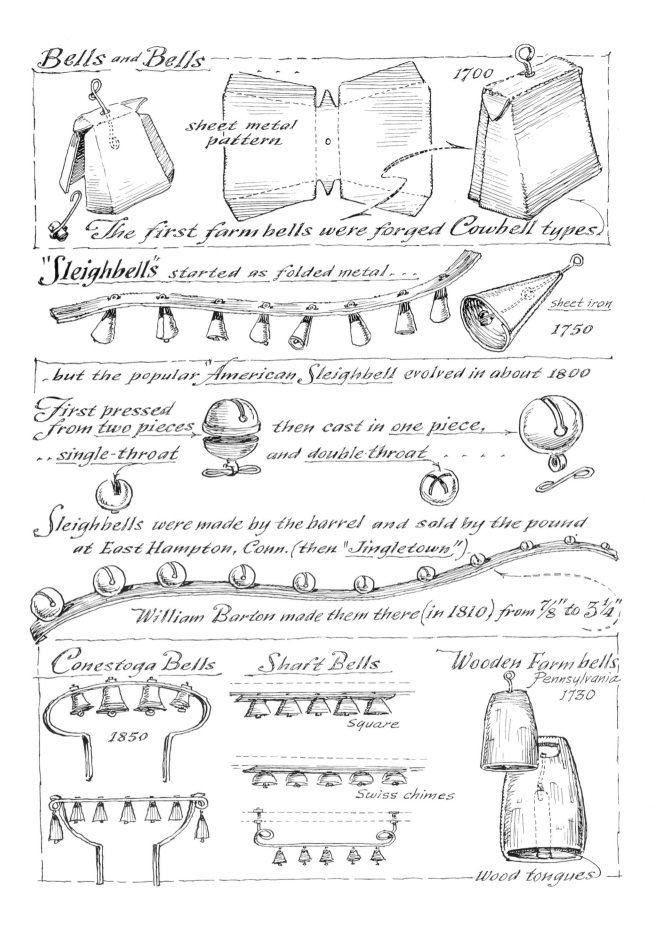

Bells and Bells

1700

sheet metal pattern

The first farm bells were forged Cowbell types.

"Sleighbells" started as folded metal . . .

sheet iron
1750

. . but the popular "American Sleighbell" evolved in about 1800

First pressed from two pieces,   then cast in one piece,

. . single-throat   and double-throat . . . . .

Sleighbells were made by the barrel and sold by the pound at East Hampton, Conn. (then "Jingletown").

William Barton made them there (in 1810) from 7/8" to 3 1/4"

Conestoga Bells
1850

Shaft Bells
Square

Swiss chimes

Wooden Farm bells
Pennsylvania
1730

Wood tongues

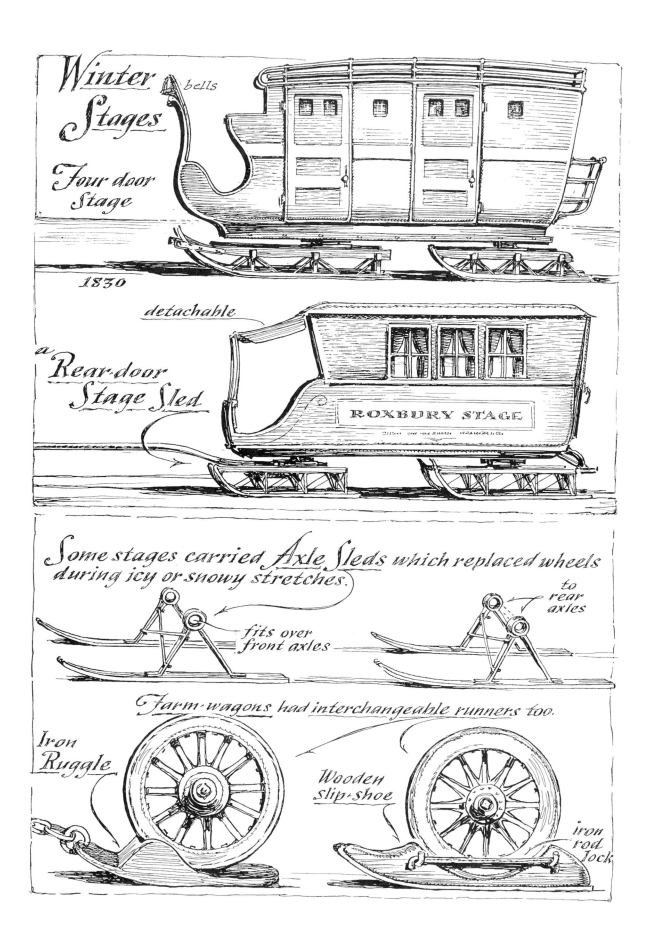

Winter Stages · bells

Four door Stage

1830

detachable

a Rear-door Stage Sled

ROXBURY STAGE

Some stages carried Axle Sleds *which replaced wheels during icy or snowy stretches.*

fits over front axles

to rear axles

Farm-wagons had interchangeable runners too.

Iron Ruggle

Wooden Slip-shoe

iron rod lock

coaches were notorious for hard-riding characteristics, but it was the stage sled that offered the easiest, most comfortable ride of the year.

The pulling capacity of a horse could be multiplied about four times if the same load he pulled in a wagon on a dirt road were placed on runners to be slid on ice. Knowing this, the early farmer had two or three sleds to every wagon, and saved all his heavy hauling for the sled season. Farm wagon-bodies were usually designed for a quick change to sled-runners. Some farmers had "wheel-sleds" which could be placed beneath chain-locked wheels. There was also a sled-runner arrangement which fastened directly over the axle, so that any wagon-body or stagecoach could have interchangeable wheels and runners.

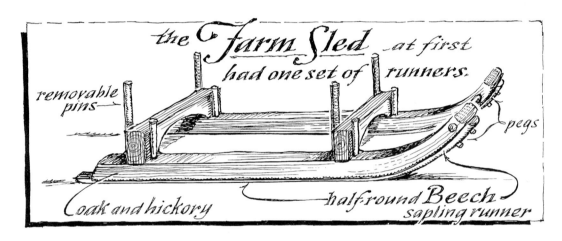

the *Farm Sled* at first had one set of runners.

removable pins —

pegs

oak and hickory — half-round Beech sapling runner

stripped for log-carrying — wagon body set on frame

The shortness of winter days was somewhat overcome by the whiteness of the snowy countryside at night, and there was more night sled traffic during winter than one might think today. Lights were seldom used, so bells were particularly necessary equipment at night. Possibly because of the nutritional deficiency of our modern eating habits, or possibly as a result of long dependence upon strong electric lighting, we seem to have much poorer night vision today than the average man had a century or two ago. The farm work and cross-country traveling

115

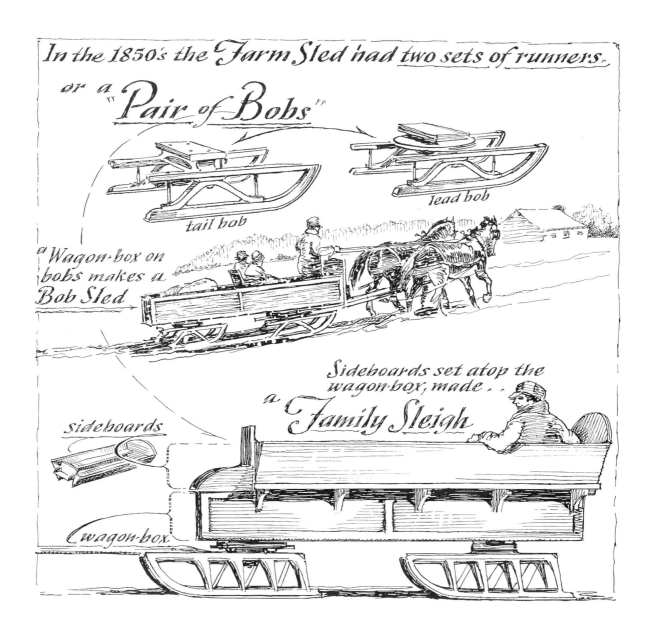

In the 1850's the Farm Sled had two sets of runners.

or a "Pair of Bobs"

tail bob

lead bob

"Wagon-box on bobs makes a Bob Sled

Sideboards set atop the wagon-box, made...

a Family Sleigh

sideboards

wagon-box

that people once did at night must often have been done in what we would consider "pitch-blackness."

December is the Season of Long Nights, and Midwinter Day (December 22) is the shortest day of the year. Winter was the season when the farmer and his family were drawn to the hearth for light as well as for heat. But the chores done by hearthlight were seldom done under stress of inconvenience; rather than being a poor light, firelight, like candle-light, could be most rich and satisfying. The hearth, like the origin of its name, was once the *heart* of the household.

116

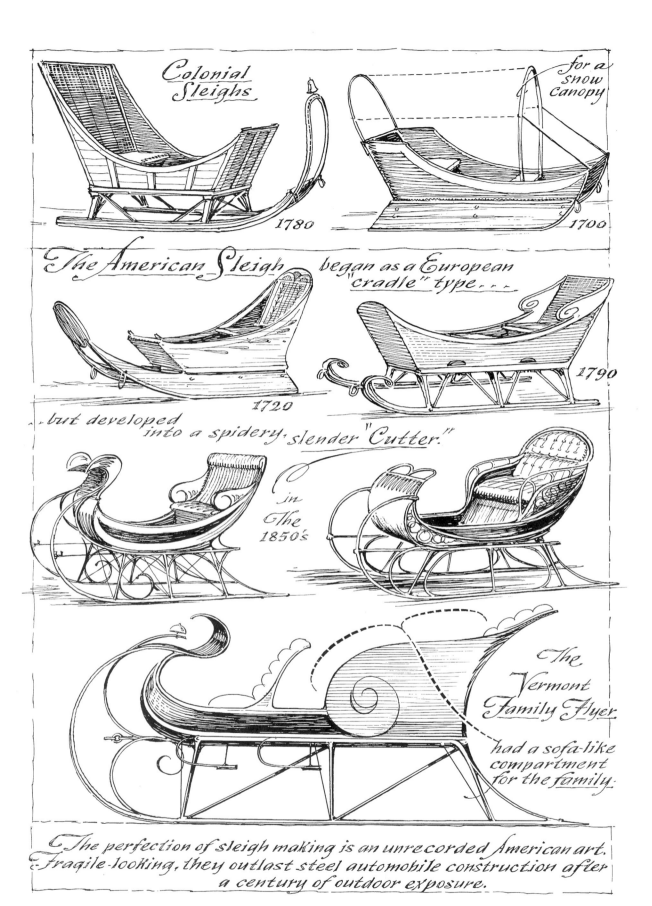

Colonial Sleighs

for a snow canopy

1780

1700

The American Sleigh began as a European "cradle" type...

1720

1790

...but developed into a spidery, slender "Cutter."

in The 1850's

The Vermont Family Flyer had a sofa-like compartment for the family.

The perfection of sleigh making is an unrecorded American art. Fragile-looking, they outlast steel automobile construction after a century of outdoor exposure.

Spinning wool by firelight, the early housewife must have presented a pleasing picture of womanhood. Exercise for the complete body, the process of wool spinning was a ballet-like movement. The woman stood at the instrument, moving back and forth to manipulate the thread in rhythm with the turning wheel; a day's work involved footwork alone equal to a five-mile walk.

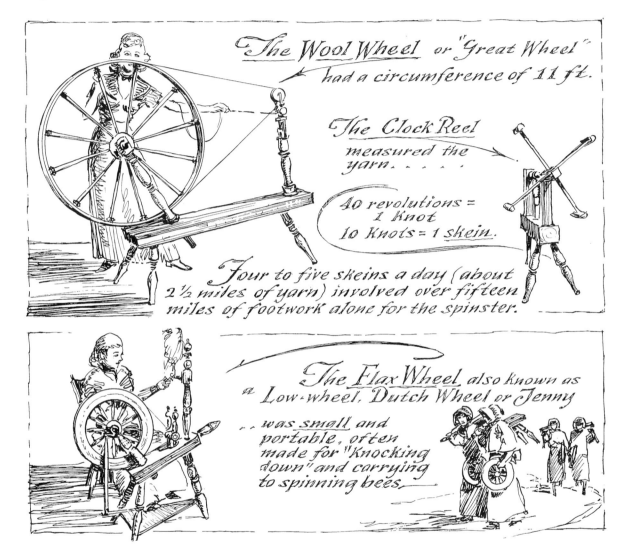

The Wool Wheel or "Great Wheel" had a circumference of 11 ft.

The Clock Reel measured the yarn . . . . .

40 revolutions = 1 Knot
10 Knots = 1 skein.

Four to five skeins a day (about 2½ miles of yarn) involved over fifteen miles of footwork alone for the spinster.

The Flax Wheel also known as a Low-wheel, Dutch Wheel or Jenny

. . was small and portable, often made for "Knocking down" and carrying to spinning bees.

The small spinning wheel that women "sat at" was not for wool but for flax. That operation too was considerable physical exercise. A day's work at a tiny flax wheel produced about a mile of thread! Any woman who could spin that much was a qualified "spinster," in the early sense of the word. The word *spinster* acquired a new meaning through the slang of schoolboys who thought of unmarried women as doing nothing but

118

sitting at home and spinning, and so a *spinster* now is an "old maid." Because women spun, and men did the weaving, women were once called the distaff keepers, or the "distaff side" of the family. The reference has lasted through the years.

That inevitable emblem of so many American roadside antique shops, the spinning wheel, has certainly stood the test of time. Made for use indoors, yet spending its second century in weather that would disintegrate any modern metal sewing machine, the old wooden spinning wheel has proved itself to be a most remarkable example of early America's craftsmanship. Many people wonder who made them. As a matter of fact, spinning wheels were most often made by experts, and our earliest wheelwrights were not makers of wagon wheels at all, but makers of spinning wheels! These were sold, farmhouse to farmhouse, from horseback. The wheelwright carried an assortment of spinning wheels which he put together on the spot as he sold them. The secret of their long life was the proper matching of seasoned woods, each one working against the other during weather changes. Without as much as one nail, square ends were "welded" into round holes, solid hard wood matched against springy wood, and a superb machine resulted which never loosened or fell apart. Probably no woodcraftsman of today would be equal to that job!

A forgotten American holiday is Forefathers' Day, which commemorated the landing of the Pilgrims on December 21, 1620. Through an error in the changing of the calendar from Old Style to New Style, however, Forefathers' Day was celebrated on December 22. Congregational Churches throughout the United States continued this holiday into the 1800's, but its nearness to Christmas lessened its observance until finally it disappeared completely from the American seasons. From a historical point of view, however, it might be good to recall the day; school children might better appreciate the scene of the Pilgrims' landing, with the grim prospect of building winter homes in New England's weather and barren countryside on a December twenty-first. Most pictures depicting this scene give the impression of balmy summer weather.

Comes Christmas!—the most profound season of the year. Christmas is so great a season that few others are given to December, even as antiquities. Yet the origination of the Christmas date is not as clear as one might imagine. For a time, the Birth was celebrated in April or May or January. Some say that the event could not have taken place in December because winter rains would have come to Palestine, closing the roads, and no flocks would have been "on the hills." But the meaning of Christmas is more clear and enduring than a mere date could make it, and few historians care to question it. Christmas is the one season that shall outlast all others.

The early American Christmas was far from what most people suppose. The Puritans banned Christmas celebrations as being sacrilegious, and as late as 1722 Samuel Sewall observed that "shops were open on Christmas, and carts came to town with wood, hoop-poles, hay and so forth as at other times."

Christmas drinks of yesterday were much richer in protein than they were in alcohol. Wassail (see recipe) is now thought of as the greatest Christmas drink of early days, but it was originally reserved for Twelfth Day (twelve days after Christmas). Twelfth Day, or Epiphany, celebrated the visit of the Magi, and whereas Christmas was more a non-drinking and religious home affair, Twelfth Day was the day set aside for visiting, drinking, and celebrating. This was the season when the punch bowl came out of the cupboard and egg-nog and wassail and sillibub became the drinks of the moment. "Nog" is an abbreviation of *noggin* (a small wooden pitcher), in which holiday drinks were once passed from mouth to mouth. Made from one block of wood, they differed from the tankard, which was staved and hooped and covered with a lid. Noggins were used at table, while tankards were reserved for hospitality at the fireside—for hot toddy and ale and mulled beers.

Sillibub (see recipe) was a southern Christmas drink of which originally only the foam was served. Its mildness made it a lady's drink, while gentlemen gathered in an adjoining room around the more alcoholic punch bowl. *Punch* was one of the earliest drinks, and the name comes from a Northern Indian word *panch*, meaning "five"; it was originally made from five ingredients: arrack, lemon, tea, sugar, and water. During the 1600's, seafaring men drank "panch," but during the next century rum was added, and the name became (much more appropriately) "punch."

Christmas spirit in a big city is often mixed with a certain essence of

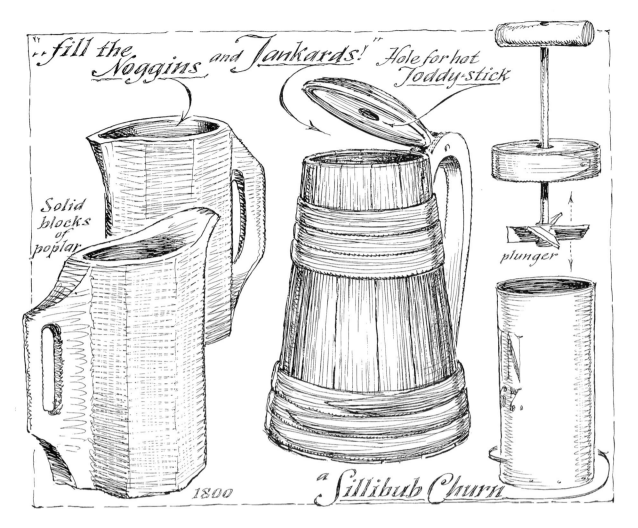

"..fill the Noggins and Tankards!" Hole for hot Toddy-stick

Solid blocks of poplar

plunger

1800

a Sillibub Churn

sadness, but in the country the peace and goodness of the season is overwhelming. It has become so entirely a season of giving that he who does not give or receive a gift seems not to partake of it at all. The countryman, however, finds Christmas in his home and in the spirit of the land, though he feels no compulsion to buy or expect perfunctory gifts. The Prince of Peace, it seems, is not too well exemplified in the big-city Christmases of merriment, excitement, big business, and almost everything imaginable but the keynote of that first Christmas, *peace*.

January

New Year's is a grown-ups' holiday, but only a century ago it was an event for children, too. People gave gifts on New Year's Day, and children went from house to house expecting a "cookie or a copper" for wishing "Happy New Year." This was the one day of the year when shops were open for trade but gifts were given to customers or visitors alike. Some customers gave hints as to what they wished for, saying, "Happy New Year—and we'll take it out in tea," or "ribbon," or some such item.

If Thanksgiving and Christmas are eating fêtes, New Year's Eve has become the season for drinking. The American custom of "seeing the New Year in," which often consists of waiting doggedly for the stroke of midnight, is something the early American wouldn't understand and many of us still are not so sure about.

Coffee for the "morning after" has been the "picker-up" for New Year's Day from the beginning, although the alcoholic New Year's Eve is something recent. Coffee was at one time a favorite New Year's Day drink. Although coffee might be termed America's national drink, we are far from being the coffee connoisseurs of a century ago, when everyone roasted and ground his own coffee. The first coffee percolator was patented in 1865, and from then on, it seems, coffee drinking deteriorated toward the present-day speedy jar-to-cup methods.

The first American restaurants were called Coffee Houses, as were our first "bars." "Having been advised to keep a coffee house," reads a 1754

Philadelphia license application, "and as some people may be desirous at times to be furnished with other liquors besides coffee, your petitioner apprehends that it is necessary to have the government license." The Exchange Coffee House of Boston was a seven-story, million-dollar building, adding in about 1820 over two hundred bedrooms.

Coffee beans were put dry into a round-bottomed iron kettle and stirred constantly for an hour or two. "If left for half a minute," the *Housewife* of 1839 says, "the kernels next to the kettle will burn and injure all the rest. Before taking up, stir in a small piece of butter and put it steaming into a closed box ready for grinding. Never under any circumstances grind more coffee than needed for one pot.

"Put a coffee-cup full," it continued, "into a pot that will hold three pints of water. Add the white of an egg, or two or three clean eggshells or a well cleansed and dried bit of fish-skin the size of a ninepence. Pour upon it boiling water and boil it for ten minutes. Then pour out a little from the spout to remove grains that may have boiled into it, and let it stand eight or ten minutes where it will keep hot *but not boil.*"

Love Feast coffee was taken in the church during the New Year's Love Feast, served with a Love Feast bun (see recipe). Love Feasts were also later held during Christmas and Easter. Coffee with a bun was served during the ceremony, symbolizing the fellowship of the church and the brotherhood of its members "making the man and his church one family." Coffee taken in this spirit of brotherly friendship took the place of the chocolate and tea which were served in the English church.

Church buns and coffee. 1800

The ice-cutting season didn't have a date. Just before winter sealed the waters across the northern states, farmers looked to their ice-houses. The pungent rotting sawdust was shoveled out and used for plant fertilizer, and a five-inch layer of new sawdust was spread to await the cutting season. The ice must be frozen to the proper thickness, and the day

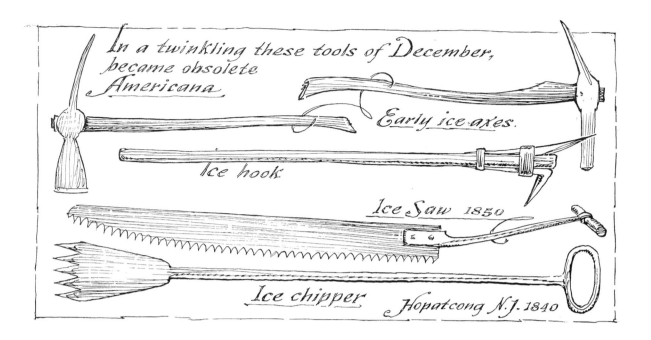

In a twinkling these tools of December, became obsolete Americana

Early ice axes.

Ice hook

Ice Saw 1850

Ice chipper   Hopatcong N.J. 1840

must be dry and windy so that the cakes won't melt and stick together. Two days or so of hard work could fill an average farm ice-house, which seems a lot more labor than just opening a refrigerator door and taking out the ice cubes. "But shucks," as one Vermonter remarked, "it takes

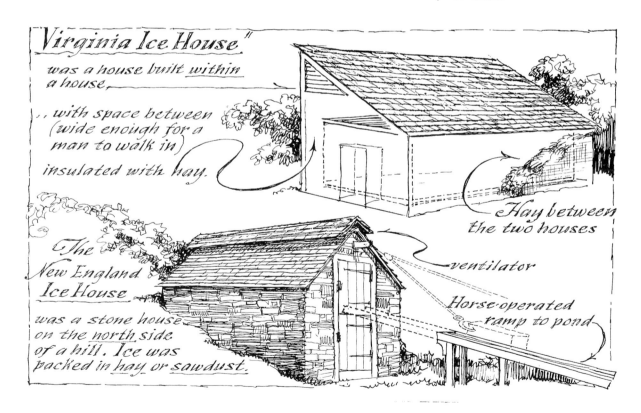

"Virginia Ice House" was a house built within a house,

.. with space between (wide enough for a man to walk in) insulated with hay.

Hay between the two houses

ventilator

Horse-operated ramp to pond

The New England Ice House was a stone house on the north side of a hill. Ice was packed in hay or sawdust.

124

more than a few days' work to buy the box and pay for the current. Besides, I appreciate ice of my own cuttin' better."

Snowshoes in the northern United States were once as common in winter as rubbers or galoshes; they were used for convenience and not for sport. The Eskimo and American Indian had skin-covered skis in the early 1800's, although they are supposed to have originated overseas; snowshoes of that early date were nearly beyond improvement. During the gold rush days in California, a part of the mail run was done on skis, and by 1860 there were ski-clubs in California and Nevada. Up until this time in the United States, however, skis were generally called snowshoes and "long snowshoes." What we now call snowshoes were then called "pads."

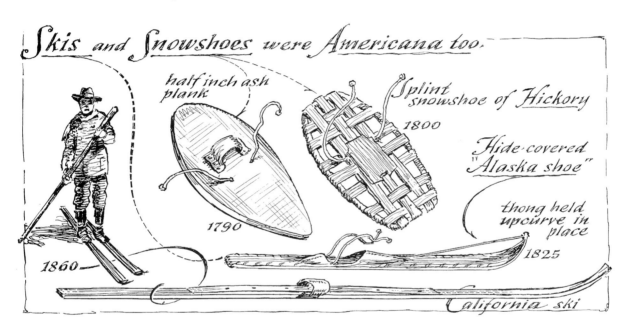

"Snowshoe Thomson" never wore a pair of snowshoes in our sense of the word, but he was the greatest American pioneer with what we call skis. His mail route was nearly a hundred miles long, his pack often a hundred pounds, and his task was to deliver mail in the mountains between Carson City, Nevada, and Placerville, California. Few modern skiers (who excel only in downhill performance) would be a match for yesterday's cross-country experts.

Some of the earliest snowshoes were oval slabs of wood (later improved into splint models). In many cases the American "ski" evolved from a sliding version of the solid wood snowshoe, for farm boys and farmers often strapped barrel-staves to their shoes to do winter chores.

January was usually the season for winter road-work in the northern states when the snow was packed and graded to make the sledding season as long as possible. Snow was shoveled into melted or otherwise bare spots by snow-wardens, fed into the covered bridges, and packed down with giant snow rollers by the road commissioner to keep the sleds going. Snow-rollers are among the rarest of antique vehicles—perhaps no more than eight exist in America. Snow roads of a century ago were nursed along during the wintertime, just as modern ski-runs are, and when March winds melted the snow of the northern countryside, most of the old-time roads were still snow-packed.

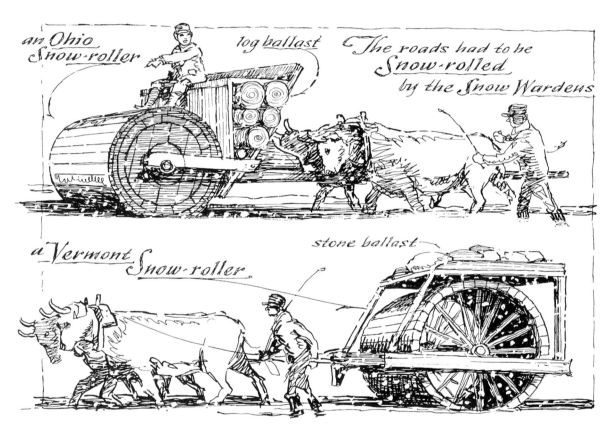

an Ohio Snow-roller — log ballast — The roads had to be Snow-rolled by the Snow Wardens

a Vermont Snow-roller — stone ballast

January was the month when the farm shop or forge-barn was at its busiest, readying all the tools for spring work. The early farmer seldom went to town for his tools or repairs, but made them right on the spot. Everything from nails to plowshares emerged from the farm shop, where the forge fire seldom went out during January.

126

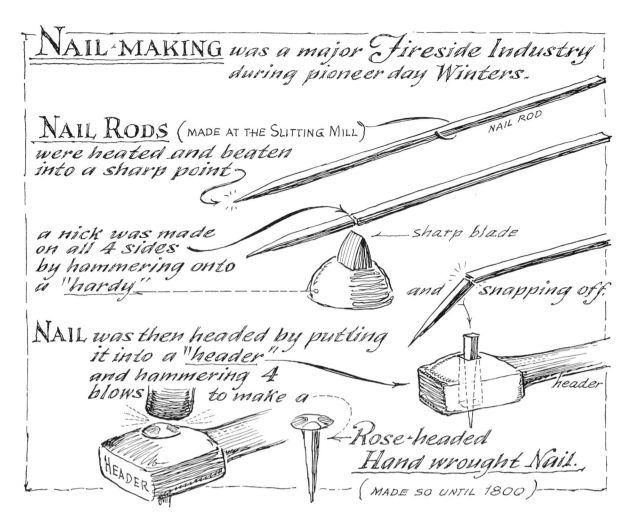

NAIL·MAKING *was a major Fireside Industry during pioneer day Winters.*

NAIL RODS (MADE AT THE SLITTING MILL) *were heated and beaten into a sharp point*

NAIL ROD

a nick was made on all 4 sides by hammering onto a "hardy"

sharp blade

and snapping off.

NAIL *was then headed by putting it into a "header" and hammering 4 blows to make a*

header

HEADER

Rose-headed Hand wrought Nail.

(MADE SO UNTIL 1800)

Plow Monday was the first day after the end of Christmas festivities, when the back-to-work spirit started with getting all farm equipment in shape. The farming season was regarded as from Plow Monday (the Monday following Epiphany on January sixth) to Lammis Day (August the first). In ancient England, plowmen went from door to door in disguise to beg gifts on Plow Monday, and there were parades of decorated plows through the villages. Early American farmers, however, were content with the date as their "Labor Day"—the signal for getting back to work.

February

All the months of the year
Curse a mildish Februeer.
OLD FARM SAYING

"A February spring," says the almanac, "ain't worth a thing." And February is often the worst season of winter, the season of drear days and howling blizzards. The word *blizzard*, as used to describe a snow-storm, is a recent Americanism, born with the big snow of 1880. Before that it had meant a "lightning storm" (a "blitzer," from the German word for lightning). A New York newspaper man, it seems, got his meteorological nomenclature confused in describing the big storm, and the new name stuck. Now any snowstorm with winds from forty miles an hour upward, a temperature of about zero, and an abundance of snow is an official blizzard.

England's Candlemas Day and America's Ground-hog Day have nothing in common except that they are both on February second. This day has always been the countryman's half-way mark in the season of winter, the day when the farmer made his mid-winter inventory.

The provident farmer on Candlemas Day,
Has half of his fires and half of his hay.

So farmers with more than half their firewood or food gone on February second did best to make necessary provision for the stark, inclement days yet ahead.

January and February have always vied for the distinction of being the coldest month of the year, or for being the "season of midwinter"; the truth, however, depends on where you live. For western and central

United States, January gets winter's crown (January 8-14) but the eastern states favor the month of February (February 5-11). Somewhere within these locations and dates you will find your "coldest day" or "coldest week."

Midsummer and midwinter were the old seasons for pruning. "Graft and prune," we were further reminded, "on the wane of moon." The old-timers pruned small trees in the summer and the larger trees in winter, often to get immediate firewood. Land leases of the early days, which prohibited the felling of trees, allowed tenants to "prune only as much underwood from the trees as could be reached *by a hook or crook*." Hence the time-worn cliché. Proper pruning, however, is not merely removing dead limbs, but a complicated art of editing nature to increase and improve fruitfulness.

Pruning tends to exercise the mind as well as the body. The steady flow of minor decisions involved is good for the soul; a man who spends a few hours in such practice immediately feels better and stronger for it. Aside from this psychological benefit, the stimulation of being out-of-doors in midsummer or midwinter makes pruning a completely fascinating farm chore. Many a grave decision concerning the farm, and even far beyond, must have been made while pruning. A passage in Leviticus suggesting that the children of Israel prune their vineyards six years in seven had connotations beyond the orchard, yet many farmers religiously adhered to it, and the idea of forgoing the pruning every seventh year reached the early farm manuals.

The proper time for wood-cutting has many variations. The "second run of sap in August" as the best time for cutting fence-railing is still disputed by many; a nation-wide vote over a century ago would have chosen February as the best month for felling fine building timber—and, of course, "during the old moon." At least during February a century or two ago, no American community was without the ringing of axes; the drawings for this month in the early almanacs invariably showed the farmer felling trees and cutting fence-posts. "Now season is good to lop or fell wood," said Tusser of February.

Most painters depicted the pioneer American woodsman with a large, flaring type of ax. Actually, the more expert the American axman, the smaller was his felling-ax. Sharp enough to use as a plane or even as a

129

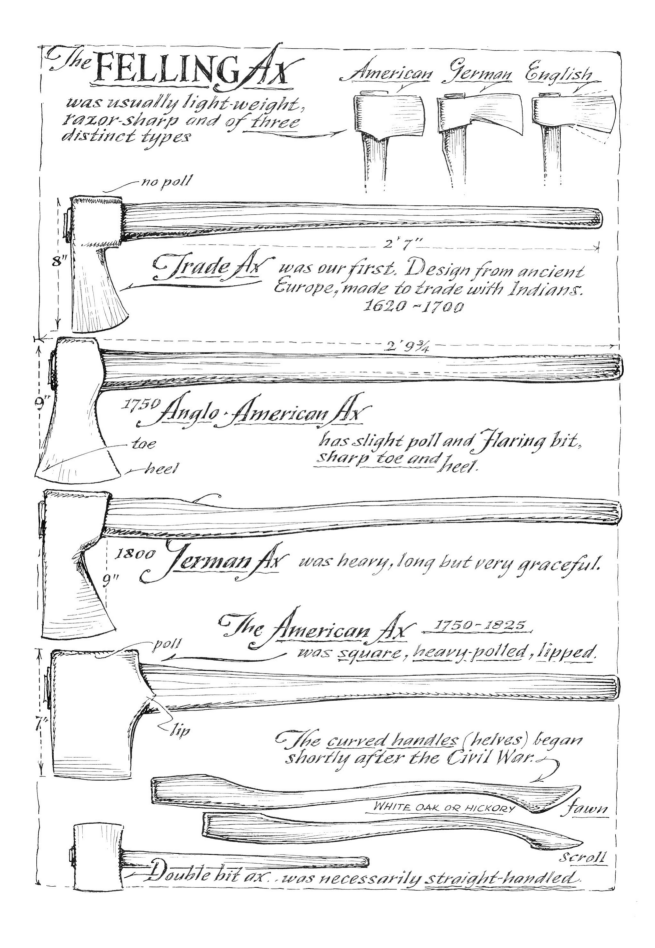

The FELLING Ax

was usually light-weight, razor-sharp and of three distinct types

American German English

no poll

8"

2' 7"

Trade Ax was our first. Design from ancient Europe, made to trade with Indians. 1620 ~ 1700

2' 9¾

9"

1750 Anglo-American Ax

toe

heel

has slight poll and Flaring bit, sharp toe and heel.

1800 German Ax was heavy, long but very graceful.

9"

The American Ax 1750 - 1825 was square, heavy-polled, lipped.

poll

7"

lip

The curved handles (helves) began shortly after the Civil War.

WHITE OAK OR HICKORY fawn

scroll

Double bit ax was necessarily straight-handled.

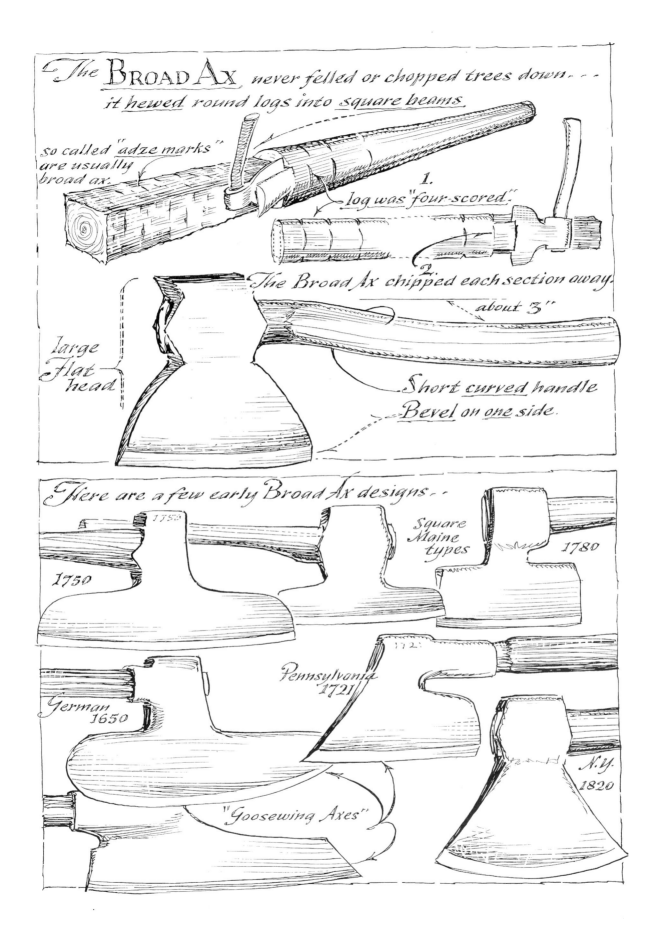

The BROAD AX never felled or chopped trees down — — it *hewed* round logs into square beams.

so called "adze marks" are usually broad ax.

1. log was "four-scored".

2. The Broad Ax chipped each section away. about 3"

large Flat head

Short *curved* handle. *Bevel* on *one* side.

Here are a few early Broad Ax designs — —

1750

Square Maine types

1780

German 1650

Pennsylvania 1721

"Goosewing Axes"

N.Y. 1820

handknife, the early felling-ax would seem ridiculously small to a person today. European ax-heads were larger, with blades flared beautifully and outweighing their poll (or butt end). American ax-heads were ugly, their blades smaller than their polls, and almost exactly square in general outline.

The illusion of the gigantic early American felling-ax is often the result of our mistaking *hewing-axes* or *broadaxes* for *felling-axes*. Broadaxes were tools for surfacing and paring, but never for felling trees. Those so-called "adzed beams" in old houses were seldom adzed, but *broad-axed*. This was the quickest way to make a square beam out of a round log; railroad ties, for example, were round logs hewed square by broadaxes. The broadax was also used for removing bark from large trees, and was sometimes called a barking-ax.

Today we usually order a cord of firewood without knowing what kind of tree it came from. Often even the seller doesn't know. A while back, however, any small child could tell you the kind of firewood by its appearance or by the smell of its smoke. A century ago there were still steamboats and commercial engines running on wood fuel, and just as there is now plain, high test, or premium gasoline at your filling station, there was a wide variety of fuel woods to choose from. At the top of the list was hickory, hard maple next, and so on down the line according to your fuel needs and your pocketbook. The soft pines which shower sparks and cause lively flames were used only for kindling, while the hardest wood was the most prized fuel.

February was the season for cleaning and scouring the soup pot. All during the winter, the farm soup pot was kept going by adding stock as fast as the soup was eaten. The fire never went out under the giant iron pot, and, somehow, neither did the flavor of the soup. Up North, any farm soup was referred to as "pot likker," but the original Southern pot likker was the concentrated liquid of any variety of cooked vegetables. Whatever was left of turnip greens, cabbage, collards, or beans was tossed into the Southern likker pot and served at all meals to be "sopped up with corn bread." Only recently, the concentrated juice of cabbage has been found to be effective in healing ulcers. With Vitamin U doing the work, the juice eases pain in a period of days, and cures have been effected within five months. This remedy was tracked down by way

"When pot likker's low
Or ceases to stew,
The farmer doth know
That the Winter is through"

of the healthy stomachs of Southern hill people and their pot-likker diets.

Being the end of the old farm year, February's last days were like today's New Year Season. Accounts and diaries were closed and inventories were made. There was talk of spring and the new farm year. Toward the South, planting had already begun; up North there was the first color of "sap-flush" in the willows, and preparations were being made for maple-sugaring. The Round of Seasons had ended, and the American farmer had finished another year of living his creed—that there is a proper time for everything on earth, and that all the earth has to offer appears in its appointed time. Such were the seasons of everyday life in America's past.

Recipes from early American Farm Days

Treenware.
Kitchenware from trees.

Cedar Dipper

Whitewood plate.
(TOP SIDE FOR DINNER
BOTTOM FOR DESSERT.)

Applewood herb Mortar

Pie crimper

Gingerbread print

Maple — Butter paddle

Cooky roller

Meat pounder
Sassafras Paddle

Masher

These interesting pieces of information are given for what they are worth, more as an insight into the old ways of living than as suggestions for your practical use. But they are all worthy of experiment (some of them you will find remarkable), and as a student of the ways of early America, you may find that they make looking into the past a richer experience.

APPLE BUTTER

For two gallons, use one bushel of peeled and quartered apples. Dump into a large kettle with just enough water to start the cooking. Cook over a hot fire (outdoors preferably). Mix one gallon of cider and three pounds of sugar, add spices (⅛ cup of allspice; ¼ cup of cloves; ½ cup of cinnamon), and boil to the consistency of molasses. Add this mixture to apples and stir continuously for about 7 hours. Then bank fire and allow to cool.

APPLE PANDOWDY

Arrange 2 cups sliced apples in bottom of buttered baking dish. Sprinkle with one-half cup molasses or brown sugar. Pour Cottage Pudding batter on top and bake.

BEER

Farm beer

A handful of hops to a pail of water, and
A half pint of molasses.
(Spruce mixed with the hops improves the taste.)

Ginger beer

1 cup ginger

1 pint molasses

1½ pails water

1 cup lively yeast

CANDIED MINT LEAVES

Pick large, pretty leaves of spearmint or peppermint; clean and dry. Dip in the whipped white of eggs. Hold by stem and coat both sides with sugar, then lay carefully on waxed paper. Allow to dry well before packing in boxes. Serve as candy or with tea. The leaves will keep sugary and green for about a year.

a Chestnut Mortar and pestle . . . for making butter, pudding, etc.

CHESTNUT PUDDING

1 cup chestnuts

3 cups chopped apple

1 cup chopped raisins

½ cup sugar

1 quart water

Peel shells off chestnuts, cover kernels with water and boil till skins can be easily peeled off. Pound chestnuts in a mortar and mix thoroughly with other ingredients. Bake until the apple is tender—about 30 minutes. Serve cold with sweet sauce.

CIDER CAKE

1½ cups sugar
1¾ cups sweet cider
¾ cup butter

4½ cups flour
1 teaspoonful each of soda,
  cinnamon and cloves

CLEANING KNIVES

Take a potato, cut in half, and dip the cut part in brick dust (cleansing powder or other mild abrasive will suffice nowadays). Rub the knife with it.

CLEANING RUGS

Scatter damp tea leaves over the rug and sweep up with a stiff broom. Cleans a rug without raising dust.

COLORED FIRES FOR FIREPLACES

Pine cones, dipped in solutions given below, will produce color. Don't mix these in your metal pans; use crockery or wooden containers. Don't leave the product within reach of children or pets; use rubber gloves while preparing.

Pick the color you want; mix its particular chemical in water, one pound to the gallon. Put pine cones in cheesecloth bags, dip them in liquid and allow them to dry well.

For:		use:	
	yellow		table salt
	purple		lithium chloride
	orange		calcium chloride
	red		strontium nitrate
	emerald		copper nitrate
	blue		copper sulfate
	vivid green		borax (or Bordeaux mixture)

COLD REMEDIES

Onions were once used as a remedy for colds and rheumatic pains, and were eaten boiled or raw twice a week as precaution. Onion-syrup cold remedy for children is made by slicing mild onions, sprinkling sugar on them, and simmering in the oven. A teaspoon of the juice of this four or five times a day was prescribed.

Lemons were and still are effective for relief of colds. An 1802 almanac says, "Roast a lemon, avoiding to burn it; when thoroughly roasted, cut into halves and squeeze juice upon 3 tablespoonfuls of powdered maple sugar. Mix; take a teaspoonful for cough or tickling throat."

COMMENCEMENT CAKE

2 pounds flour	1 gill yeast
1¼ pounds sugar	1 teaspoon cinnamon
1 pound butter	½ teaspoon clove or some mace
1 pint milk	1 pound raisins
½ cup wine	1 nutmeg
4 eggs	½ pound currants

Make up the flour, yeast, and milk as for bread; when fully light, add all other ingredients except raisins and currants. Pour mixture into deep pan and if weather is cool, let it stand for a day. When mixture is again very light, add the raisins and currants and bake for two hours. Commencement cake keeps extremely well.

CORN BREAD (Virginia 1800)

Dissolve tablespoon butter in 3½ pints boiling milk. In this, scald one quart of Indian meal (corn meal). When cool add ½ pint of wheat flour, a little sugar, a teaspoonful of salt, and 2 well-beaten eggs. Mix well together. Bake in two cake tins well covered.

CURRANT WINE

1 quart currant juice
2 quarts cold water
3 pounds brown sugar

Put in a cask with the bung out to ferment; when the sound of fermentation ceases, make the cask tight. Leave for one year and then bottle.

DANDELION WINE

3 pounds dandelion blooms (no stems)
3 oranges

3 lemons
3½ pounds sugar
1 cake yeast

Boil dandelion blooms, oranges, and lemons in two gallons water for 20 minutes. Let stand overnight, then strain and add sugar and yeast. Pour into a crock and cover with a cloth. Let stand for two weeks and bottle.

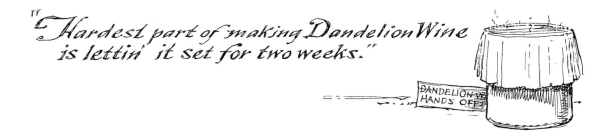

"Hardest part of making Dandelion Wine is lettin' it set for two weeks."

DYE

Red, blue, green, or *yellow dye* (New York 1760)

After boiling cloth, bones, or feathers in alum-water, steep in:
 infusion of redwood for red
 blue-pot or juice of elderberries for blue
 lime-water and vertigris (or nitrate of copper) for green
 tincture of saffron for yellow

Slate or *cream dye*

Teagrounds boiled in iron and set with copperas makes a good slate color. Cloth boiled in heavy tea will have a rich cream color.

Purple slate dye

Rusty nails (or any rusty iron) boiled in vinegar with a small bit of alum makes a fine purple slate color for cloth or wood.

Brown dye

The scaly moss from rocks makes good brown dye. Gather moss and boil it in water for 3 or 4 hours. Skim out moss and put in goods, and boil until color is correct. It will never fade.

Buff dye

Fill kettle with small pieces of white birch bark and water. Steep for 24 hours, do not boil; skim out bark. Wet cloth in soapsuds, then put in the dye and stir well, airing often. When dark enough, dry, then wash in suds. It will never fade.

ELECTION DAY CAKE

½ cup butter	½ teaspoon soda
1 cup white bread dough	1 teaspoon cinnamon
1 well-beaten egg	¼ teaspoon clove
1 cup brown sugar	¼ teaspoon mace
½ cup raisins and figs, chopped	¼ teaspoon nutmeg
1¼ cups flour	1 teaspoon salt

Work butter into dough with hands, add egg, sugar, milk to moisten, fruit, dredged with two tablespoons flour, and the rest of the flour mixed with remaining ingredients. Put in buttered bread pan, cover, and let rise 1¼ hours. Bake 1 hour in slow oven (300 degrees). Cover with frosting.

HAYMAKERS' SWITCHEL

1 cup brown sugar	¾ cup vinegar
½ teaspoon ginger	2 quarts water
½ cup molasses	

Mix and chill for farm hands.

HEADACHE CURES

Dr. Chase's famous old doctor book blamed tea and coffee as the cause of many sick headaches. He suggests substituting milk, or simply taking less tea and coffee.

For *headache from heartburn*, a teaspoonful of bicarbonate of soda (baking soda) in 3 or 4 tablespoonfuls of peppermint or cinnamon water, with ½ teaspoon of ginger, taken after eating.

For *headache caused by eye-strain*, exercise the eyes outward by instantaneous staring at two outward objects (to give the opposite effect of cross-eyes). Five or ten tries at this will often give instant relief.

HICCOUGHS

Just gaze steadfastly into the returned gaze of the hiccougher and he will cease hiccoughing.

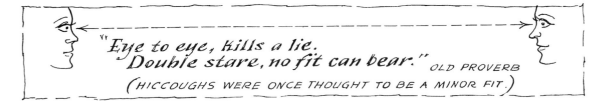

"Eye to eye, Kills a lie.
Double stare, no fit can bear." OLD PROVERB
(HICCOUGHS WERE ONCE THOUGHT TO BE A MINOR FIT.)

INDIAN PUDDING (Vermont)

Into a quart of lightly boiling milk, scald ten tablespoons of Indian meal (corn meal). Cool and add a teacup-full of molasses, butter the size of an egg, a teaspoonful of salt, also of ginger and cinnamon. Bake in a pudding dish from one to two hours. Serve with heavy cream or sauce.

LOVE FEAST BUN

4½ cups sugar
4 eggs
1 heaping tablespoon salt
1 cup butter & lard mixed

1 cup warm mashed potato
3 cakes yeast
2 gallons flour
lukewarm water

Add other ingredients (except water) to the beaten eggs, adding water last to make soft dough. Knead and keep warm until "light." Make into buns about size of a small cup and bake on greased tins.

MEAD

There are three kinds of the honey-and-water beverage, mead.

Simple mead (unfermented)

3 parts water—1 part honey

Boil over slow fire until a third evaporates. Skim, cool, pour in cask. Drinkable in 3 or 4 days.

Compound mead (unfermented)

½ lb. seeded raisins
2 quarts water
6 pounds honey

1 slice toast steeped in beer
1 oz. salt of tartar
1 glass brandy

Boil first three ingredients until raisins are soft, and half of liquid evaporates. Strain through linen; add to the simple mead at boiling stage. Add 1 slice toast steeped in beer. Skim and cool, then add 1 oz. salt of tartar dissolved in a glass of brandy (lemon peel, cinnamon, or the syrup of berries or aromatic flowers may be added here for variety).

Fill cask to top so working will spill froth over edge. Add mead to keep full until working ceases, then bung and store for 2 or 3 months.

Vinous mead or "Miod" (fermented)

Soak berries (cherries, strawberries, gooseberries, and mulberries—one kind or a combination) in water for several days. Put in cask; add honey (plain or kneaded with flour of sesame) and a slice of beer-soaked bread. Keep cask in warm room. Fermentation begins in 6 to 8 days and lasts 6 weeks. Vinous mead is then drinkable, but improves with age.

MINT VINEGAR

1 quart cider vinegar
1 pint spearmint leaves and stem tips
1 cup granulated sugar

Bring vinegar to boil. Add sugar and mint. Boil for a few minutes while stirring and crushing. Strain and bottle promptly. Mint vinegar is good in fruit punches and iced tea and a fine base for mint sauce.

ODORS FROM COOKING

Odors from boiling ham, cabbage, etc., are eliminated by throwing whole red peppers into the pot, at the same time improving the flavor.

PAINT

Inside white paint for barns

1 gallon skimmed milk
1 pound lime
8 ounces linseed oil (or neatsfoot hoof-glue)

Outside red paint

6 pounds Venetian red (35 percent sesquioxide of iron ground in oil)
1 pound resin
4 gallons raw linseed oil

White fence paint

1 gallon skimmed milk
10 pounds lime
1 pound salt

Permanent white paint

1 bushel well-burned white lime unslaked
20 pounds Spanish whiting
17 pounds rock salt
12 pounds brown sugar

Slake the lime, sifting out lumps, and mix to a good whitewash with 30 gallons of water. Then add the other ingredients and stir. Two or three coats should make a permanent covering that will not rub or wash off.

White sorghum paint

2 quarts sorghum molasses
50 pounds lime
4 pounds table salt

5 ounces alum
10 gallons water

For outbuildings, to protect wood and reduce summer heat.

PICKLE RECIPES

Quick Dill Pickles

Small cucumbers
Stalks of dill with flowers

1 quart vinegar
2 quarts water
1 cup salt

Pack washed cucumbers in quart jars with one dill stalk at bottom and one at top of each jar. Bring vinegar, water, and salt to a boil; while boiling hard, fill jars to overflowing and seal.

Quick Sweet Pickles

7 pounds yellow-ripe cucumbers
3 pounds sugar
1 pint vinegar

4 sticks cinnamon
6 whole cloves

Cut cucumbers in sticks; simmer in salted water 5 minutes and drain. Make syrup of other ingredients. Put cucumbers in syrup and boil till tender yet still firm. Fill jars with cucumber sticks; cover with syrup and seal.

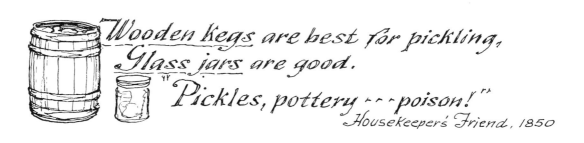

Wooden kegs are best for pickling,
Glass jars are good.
"Pickles, pottery ‑ ‑ poison!"
Housekeeper's Friend, 1850

RUST PREVENTATIVE

For grates and stoves during summer, or tools during winter, melt 3 parts lard and 1 of resin powder. Apply a thin coating with brush. Also good for brass, copper, and steel.

SALT LOOSENER

A tiny block of soft pine wood kept in the salt box will keep it loose and flowing during damp weather. Salt boxes made of soft pine also keep salt in good condition.

SILLIBUB

1 cup wine	2 cups milk
1 quart cream	1 cup sugar

Mix ingredients and beat well. Serve only the foam, sprinkled with nutmeg. (Some old recipes substitute beaten egg whites for the cream.)

SIMNEL CAKE

¼ pound butter	2 ounces candy peel and/or citron
¼ pound sugar	½ pound currants
¼ pound flour	2 eggs

Cream butter and mix in sugar. Beat eggs one at a time; add each slowly. Dust fruit with flour; add remaining flour and then flour-dusted fruit. Place in greased pan and bake at 350 degrees. Cover with simple white icing. Traditionally, icing is decorated with candied violets. Take freshly picked violets, wash, and dip in syrup of sugar and water.

SOOTHING POWDER

The powdered-wood dust of the powder post beetle is famous for its fine soothing qualities. It was once prized as a baby powder. (The powder post beetle is the one that makes the "worm-holes in ancient wood"; the tannish-colored powder falls in piles wherever the beetle is working.)

SORE THROAT

Make paper tube, put bicarbonate of soda inside. Aim at tonsil or other sore spot and have another blow the powder at it. Once every hour or two until relieved.

STYPTIC

There is no better styptic than the powder from the common puff-ball fungus. Severe bleedings at the nose have been stopped by this dry powder.

TOOTHACHE

Put powdered cloves (or oil of cloves) directly into the cavity. (This is still a seemingly miraculous remedy.)

TOOTHPOWDER

Powdered charcoal is the best deodorizer known, and one of the very best

Early Dentist sign featured a Clove —

BECAUSE CLOVES LOOK LIKE NAILS, THEY WERE CALLED CLOVE FROM THE LATIN FOR NAIL

and in the 1800's

DR. BELDING PAINLESS PULLING

it became a tooth.

cleaners. As a tooth cleaner it is temporarily messy, but is still unmatched in its efficiency as a whitener.

WASSAIL

Grated nutmeg
2 cloves
1 teaspoon ginger
Blade of mace
1 teaspoon allspice

2 teaspoons cinnamon
2 bottles of wine (claret, sauterne, or burgundy)
1 pound sugar
12 eggs
12 *hot baked apples*

Put spices in ½ cup water and bring to boil. Add the wine and heat. Add sugar. Beat separately yolks and whites of eggs. Fold yolks into beaten whites and put in punch bowl. Pour the warm spiced wine over eggs and beat until frothy. While liquid is foaming, add the hot baked apples.

WATERMELON CAKE

White Part:

2 cups white sugar	8 egg whites
1 cup butter	2 teaspoons cream of tartar
1 cup sweet milk	1 teaspoon soda
3½ cups flour	

Dissolve soda in a little warm water; sift cream of tartar in flour; mix.

Red Part:

1 cup red sugar (kept by confectioners)	4 egg whites
⅓ cup sweet milk	1 teaspoon cream of tartar
½ cup butter	½ teaspoon soda
2 cups flour	1 cup raisins

Mix above. Put red part in center and white part around outside. Desired effect easier if two people fill, one putting in the white batter, one the red.

WATERMELON RIND PICKLE

4 pounds watermelon rind (with green and pink removed)	4 pounds sugar
	2 tablespoons whole cloves
2 quarts salt water	2 tablespoons whole allspice
2 quarts vinegar	12-inch pieces stick cinnamon
1 pint water	

Cut rind in inch cubes; soak overnight in salt water (¼ cup salt to 2 quarts cold water). Drain, cover with fresh cold water, and boil until tender. Mix vinegar, water, sugar, and the spices (tied in a cheesecloth bag). Boil 5 minutes. Add the drained watermelon rind and boil gently until clear. Put in hot jars and seal.

WOOD STAIN : Early American Fine Brown Wood Stain

1 gallon skimmed milk
About 5 pounds of lime
½ pound powdered alum (or salt
   if alum is unobtainable)

Boil with walnut or butternut shucks. Strain and dilute with skimmed milk to desired color.